DEEP FREEZE COOKBOOK

Published by Paul Hamlyn Pty. Ltd.,
176 South Creek Rd, Dee Why West,
New South Wales, 2099
First Published 1971
Copyright Paul Hamlyn Pty. Ltd. 1971
Printed by Lee Fung, Hong Kong.
Registry number 0600 07086 7

DEEP FREEZE COOKBOOK

NORMA McCULLOCH

EDITOR, ANNE MARSHALL.
PHOTOGRAPHER, SVEND BENDTSEN. DESIGNER, HUGH McLEOD.

HG

PAUL HAMLYN
SYDNEY . LONDON . NEW YORK . TORONTO

PREFACE

Today most of us live at quite a hectic pace. It is an age of speed in which many a housewife is involved. The deep freeze is a blessing to her. It is an efficient low temperature storage cabinet for the wide range of frozen foods available in the supermarkets. It is also ideal for home freezing of food. Many a proud homemaker has already discovered the joys of owning a deep freeze. I hope this book will inspire them to use their freezer to its maximum potential.

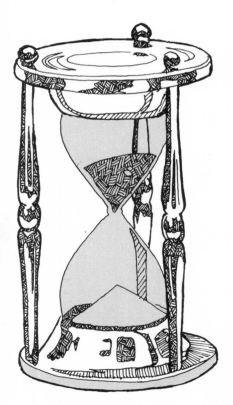

CONTENTS

INTRODUCTION

I am of the opinion that we housewives work
harder and longer at household chores than any
other occupation you care to think about. There
are not many careers which start the minute you get
out of bed in the morning, and carry on practically
non-stop all day until you wearily lower your bones
and aching feet into bed at night. Seven days a
week, fifty-two weeks a year, cooking, cleaning,
organising and budgeting, to mention just a few
of the tasks we housewives tackle each day.
However, the rewards for our sixteen hour day are .
usually the type of rewards money cannot buy,
like the pleasure of preparing a delicious meal to be
enjoyed by the whole family, the reward of
watching our children grow strong and healthy,
day by day, the sense of achievement when your
husband goes off to work on time and the children
line up for a goodbye kiss as they go off to school
looking like shining angels. Oh yes, we have our
rewards!

Next time the housework is done and you put your
feet up for five minutes' rest, I would like you to
join me in a moment's thoughtful appreciation of all
the designers and engineers throughout the world
who have tried to make our busy lives easier. Over
the past fifty years they have given us the
refrigerator, the vacuum cleaner, the washing
machine, the electric cooker, the gas cooker, the
dish washer, the clothes dryer and now the greatest
home help of all, the 'deep freeze'. No other home
appliance can give you what a deep freeze can;
the economy of saving hard earned cash by buying

food in bulk and freezing it when prices are lowest, and the convenience of always having a variety of foods on hand ready to be used whenever needed.

I sincerely hope this book will help you to enjoy the many benefits of a deep freeze.

Norma McCulloch

GUIDE TO WEIGHTS AND MEASURES

The weights and fluid measures used throughout this book refer to those of THE STANDARDS ASSOCIATION OF AUSTRALIA. All spoon measurements are level unless otherwise stated. When a recipe calls for a 'good' tablespoon, measure generously. For a 'scant' tablespoon, measure conservatively. A good set of scales, a graduated Australian Standard measuring cup and a set of Australian Standard measuring spoons will be most helpful. These are available at leading hardware stores.

The Australian Standard measuring cup has a capacity of 8 fluid ounces.

The Australian Standard tablespoon has a capacity of 20 millilitres.

The Australian Standard teaspoon has a capacity of 5 millilitres.

The British Imperial pint (used in Australia and New Zealand and in this book) has a volume of 20 fluid ounces. e.g.

1 pint (Australian or New Zealand). $2\frac{1}{2}$ cups
$\frac{1}{2}$ pint (Australian or New Zealand). $1\frac{1}{4}$ cups
$\frac{1}{4}$ pint (Australian or New Zealand).   5 fluid ounces

IMPORTANT POINTS
AMERICAN CANADIAN WEIGHTS

AMERICAN weights and measures are the same except for the tablespoon.

Housewives in AMERICA and CANADA using this book should remember that the AUSTRALIAN standard measuring tablespoon has a capacity of 20 millilitres, whereas the AMERICAN/CANADIAN standard measuring tablespoon has a capacity of 15 millilitres, therefore all tablespoon measures should be taken generously in AMERICA and CANADA.

It is also important to know that the imperial pint (20 fluid ounces) is used in Australia, whereas the AMERICAN/CANADIAN pint has a volume of 16 fluid ounces.

OVEN TEMPERATURES

This is an approximate guide only. Different makes of stoves vary and even the same make of stove can give slightly different individual results at the same temperature. If in doubt with your particular stove, do refer to your own manufacturer's temperature chart. It is impossible in a general book to be exact for every stove, but the following is a good average guide in every case.

Description of Oven	Automatic Electric	Gas Thermostat Setting °F	Gas No.
Cool or low	200	200	0-$\frac{1}{2}$
Very slow	250	250	$\frac{1}{2}$-1
Slow	300-325	300	1-2
Moderately slow	325-350	325	3
Moderate	350-375	350	4
Moderately hot	375-400	375	5-6
Hot	400-450	400	6-7
Very hot	450-500	450	8-9

HAPPINESS IS HAVING A DEEP FREEZE

The advantages of owning a deep freeze are numerous. For example the budget-minded housewife can buy food in bulk at wholesale prices e.g. meat, fish, poultry, vegetables and fruit. One shopping trip can take the place of several, which also saves money. All types of meals can be prepared, then stored in your freezer, ready for unexpected guests, or for the odd occasion when you are not feeling well enough to cook. Just think of the advantages a freezer can offer; planning a party, preparing school lunches in advance, meals in a hurry, serving foods out of season, preparing food for Christmas, Easter and holiday times, or simply preserving left-over foods to save waste.

FOR THE FARMER'S WIFE

My neighbours are farming people who often have a whole carcase to cut up and freeze. They also have access to a whole crop of vegetables and vegetables mature all at once. It is no problem to preserve them when you own a deep freeze (see pages 28-31). If you are a farmer's wife or the mother of a large family, it is false economy to buy a small freezer when a larger freezer enables such saving as freezing a whole year's supply of meat, vegetables and fruit. It is important to remember that bulk quantities of food should be frozen only when there is a light load in the freezer, so that it will continue to function efficiently and the quality of the frozen food will be retained.

There is also the added convenience of being able to cater for extra hands at shearing time with huge supplies of frozen pies, scones, biscuits and complete meals such as cottage pie, braised chops, curried sausages etc. Several of my farming friends have even bought a second large freezer. Town folk will probably find this statement hard to believe, but I can assure you it is true.

FOR THE WORKING WIFE

For the housewife who goes out to work every day a deep freeze is a great helping hand. You will probably find the small or middle-sized freezer large enough, but the choice depends on the size of family, and whether you intend buying food in bulk. However, your greatest advantage in owning a deep freeze is in preparing family meals, which can be frozen ready to be taken out of the freezer, reheated and served quickly. For example, the next time you prepare a casserole or a curry, cook a triple quantity, use what you need for the evening meal, then divide the remainder into separate aluminium foil or tinfoil dishes or suitable freezer containers, seal with freezer tape, label with date and contents then freeze. As an alternative container, a large plastic bag placed in a casserole as a lining is ideal. Once the food is frozen, remove the plastic bag and contents from the casserole, extract the air from the plastic bag with a vacuum pump (see pages 15 and 104) and seal with a wire twist. When required, the block of frozen food may be removed easily from the plastic bag, placed back into the original casserole, left to thaw for half to one hour in a warm place and then placed in a hot oven (400 °F. gas No. 6-7) until heated through (approximately one hour).

Dozens of chops, fillets of steak and beefburgers,

for evening and weekend meals, can be frozen ready to take straight from the freezer and put into the frypan or onto the barbecue (see page 22). Packed school lunches can be prepared on a Sunday to cater for the whole week. Remember to omit salad vegetables and always package sandwiches well (see page 104).

FOR THE YOUNG BUSY MOTHER
If you are a young sophisticated housewife, chained to the house with babies and toddlers, you will probably find the greatest relaxation and joy in entertaining friends and relatives to a delicious evening meal. When the children are tucked up safely in bed you can unwind by serving a meal which will leave your visitors wide-eyed. With a little forethought and your deep freeze, you can serve out of season delicacies such as asparagus, mushrooms, salmon, crayfish, strawberries and pawpaw etc. Ingredients for exotic meals can be frozen, when in season, ready to be used occasionally throughout the whole year.

Several types of canapés and savoury dips can be prepared and frozen. Freeze canapés on flat smooth trays and, when frozen, pack into airtight containers. Make full use of your freezer to prevent being rushed on special occasions.

Prepare time consuming sauces in bulk. Separate the sauce into meal size quantities and freeze in plastic or wax-coated tubs. Remember two points:

(1) Use rice flour for the thickening of sauces, as ordinary flour causes sauce to separate on thawing.
(2) Liquid type foods expand when frozen, so always leave approximately 1-inch head room for expansion.

Feeding babies and toddlers can be a problem.
Why not let your freezer help you here too?
Prepare large quantities of assorted purées of
vegetables, meat, fish and fruit. Deep freeze small
quantities in ice cube trays. Once frozen, tip cubes
into clearly labelled plastic or polythene bags,
remove the air with a vacuum pump and seal with a
wire twist. Reheat as many ice cubes as required
in the top of a double boiler.

As you can now see, there are many advantages in
having a deep freeze, so have a little happiness
yourself and treat yourself to one!

ICE COLD FACTS

When food is deep frozen the juices, flavour, colour and nutrition content are retained and held over long periods of time. The quality of the food on thawing should be practically the same as when first frozen. It is at zero temperature that most of the enzyme action is arrested, which prevents food deteriorating. The lower the temperature reached and maintained the longer the food will be safe from deterioration. Food spoilage is therefore delayed when kept below zero (see Storage Time Guide). However, the action of deep freezing itself can cause deterioration in three ways:

(1) If food is not protected from freezing air, by good quality airtight packaging materials, it will dehydrate or desiccate, losing valuable juices, flavour, colour and nutriment content, because the action of freezing air is to draw warmth and moisture from the food, consequently removing goodness also. Therefore, good airtight packaging materials are vital to successful frozen foods.

(2) If air is present in the package of food, oxygen is absorbed into the product, causing tainted flavours in the food. Proper packaging in moisture-vapour proof materials or a container free from air reduces oxidation to a minimum.

(3) It is a fact that deep freezing has a natural tenderising effect on all types of food, because ice crystals form inside the food, breaking down fibres. To prevent mushy foods, caution must be observed when freezing cooked dishes containing vegetables, e.g. casseroles, stews etc. Always leave adding the vegetables to such dishes until practically cooked.

I recommend approximately ten to fifteen minutes before completion of cooking time. The natural deep freeze tenderising effect will complete the softening.

Your deep freeze is a home help. It is up to you to learn the best methods of making full use of it. World-wide statistics have proved that frozen foods retain far more vitamins and nutritive value than any other method of preserving. However, always remember the importance of hygiene. Clean hands and utensils are a must when handling food.

Air is one of food's worst enemies. Always vacuum seal whenever possible, a vacuum pump especially designed for deep freezing is reasonably priced and can be purchased from any good kitchenware store. Another method of removing the air before sealing is to simply press air out of packages with your hands before tying off with a wire twist.

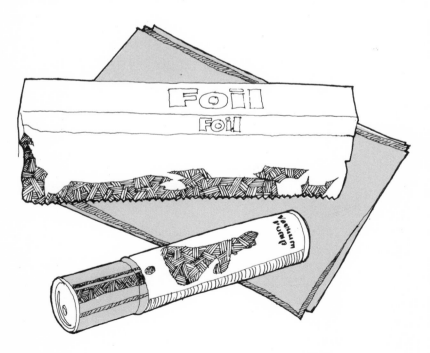

THE CHOICE IS YOURS

THE CHEST FREEZER

One of the most popular home freezers appears to
be the 'CHEST' type which is available in several
sizes. The 6 cu ft freezer holds approximately 200 lb
of frozen foods if stacked correctly. The 10 cu ft size
will hold approximately 350 lb. Both these sizes
will suit the smaller family. They keep everything
fresh no matter how high the temperature outside,
and give easy access to food. However, the larger
family or farmer's wife would probably need the
16 cu ft freezer which holds 550 lb or the 22 cu ft
size holding around 800 lb. Foodstuff may be kept
frozen all year round in the large chest freezer which
makes them economical for country properties.

The chest type freezer usually has a magnetic seal
around the complete lid, which prevents warm air
getting into your freezer. The quick freeze area is
situated over the electric motor end, however, good
quality freezers have the freezing pipes all around
the walls and at the bottom, not just at one end.
Vinyl coated dividers are provided with the larger

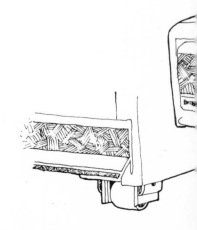

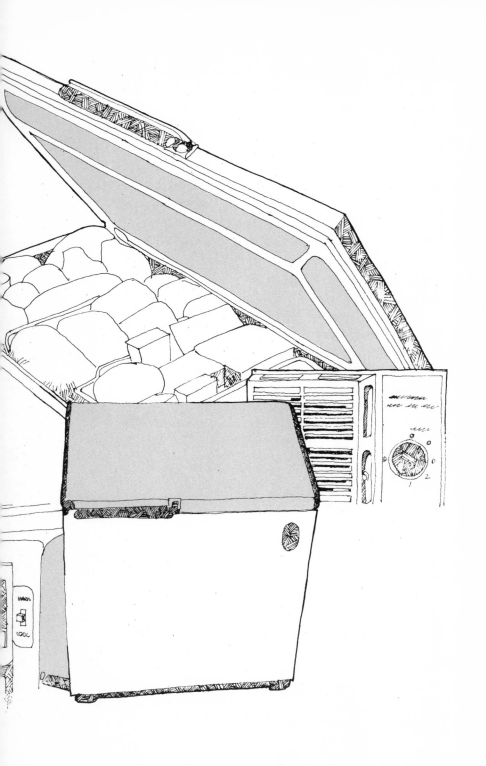

chest freezers. The dividers enable you to keep different types of food separate. Vinyl coated baskets are also supplied with the chest type freezers. These are ideal for an assortment of small items.

Most well made chest type freezers have a temperature control situated on the outside, which enables you to switch to quick freeze when putting a large quantity of food in at the one time. Any type of food which is frozen quickly is of far better quality when eaten later.

THE UPRIGHT FREEZER

If kitchen space is a problem I would strongly recommend the 'UPRIGHT' or 'VERTICAL' freezer as it is a great space saver, taking up very little valuable floor space. Once again, there are several sizes to choose from. The top quality upright freezer usually has vinyl coated adjustable shelves and steel interior lining, with freezing pipes situated all round, lining sides, back, bottom and top i.e. inside the lining, out of sight.

The quick freeze area is situated at the bottom directly over the freezer motor. A temperature control and quick freeze switch is also usually positioned at the bottom. With a touch of the toe you can switch to quick freeze. Once food is frozen switch back to thermostatic control. Remember the quicker you freeze food the better the quality. There is usually plenty of door space for easy access to smaller items. An ice bucket and several ice cube trays are also usually supplied. Once again, a magnetic door seal gives a perfect seal every time. No chance of warm air getting in which means no waste of precious power. Of course, the convenience of the upright freezer should not be forgotten for when the shelves are full of frozen food, it gives you a supermarket in your own kitchen!

THE REFRIGERATOR FREEZER

If you need a new refrigerator and have never owned a deep freeze, I would suggest you trade in your old refrigerator for a new two door dual temperature 'REFRIGERATOR-FREEZER'. This is two appliances in one, a household refrigerator plus a true long term deep freeze cabinet. You will never regret the extra cost because you will have the pleasure of automatic frost-free in both compartments. Frigid fresh air is injected by a fan into the refrigerator and freezer and circulates continuously so there is no frost or condensation. Defrosting is unnecessary and you can identify your frozen food immediately because it does not ice up. You can buy frozen food specials at the supermarket, store them in your freezer and save several dollars a month on your grocery bill. You will also find your first deep freeze a great convenience as left-over foods can be packaged, frozen and used weeks or months later.

The deep freeze storage cabinet of the refrigerator-freezer is usually positioned on top of the refrigerator cabinet. In some models they are side by side, and a few have been manufactured with the freezer under the refrigerator compartment. All models are usually fitted with an interior light and vinyl coated adjustable shelves. In the refrigerator compartment there is an egg rack, a dairy chest for butter and cheese and a crisper for vegetables. Once again, magnetic door seals are fitted to all quality refrigerator-freezers to keep cold air in and warm air out. The dual refrigerator-freezer takes up very little precious kitchen floor space. You will never regret buying a good quality deep freeze.

FREEZING MEAT

Deep freezing meat, poultry and game is easy, economical, and very convenient. ATTENTION FARMERS— do not make the mistake of freezing meat too soon after butchering, a mistake which causes the toughening of meat. Always hang BEEF and LARGE WILD GAME carcases in a chiller or refrigerator for approximately 6-10 days before cutting up, packaging and freezing. Hang PORK and VEAL for approximately 24-36 hours. MUTTON—always allow 2-4 days and LAMB 1-2 days. Hanging meat before packaging to freeze always improves the texture and flavour.

IMPORTANT POINTS
(1) Always use good quality packaging materials to prevent meat from drying out (see page 105).
(2) If large quantities of meat are to be stored for long periods, I recommend the addition of mutton cloth covers tied firmly over the first wrapping, which will prevent damage to the airtight seal. Damage can be caused through the heavy solid blocks of meat banging together when a particular package is being located.

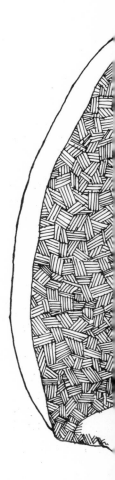

(3) Labelling is a must, if chaos is to be avoided, for many items look alike when frozen. Always label clearly with the contents, date and quantity.

To prevent STEAKS and CHOPS sticking together when freezing, place a double layer of plastic or waxed paper or a single layer of aluminium foil between each item. Place into a strong plastic or polythene bag and remove the air before sealing and freezing. An alternative method is simply to place chops and steaks onto a smooth flat tray, place tray into the freezer without a cover and freeze for approximately 2 hours. Once frozen, remove from tray and package into strong plastic or polythene bags. Remove the air and seal tightly, then return to freezer. Items frozen by this method before packaging will not stick together—a great advantage to the busy housewife, as it is not necessary to thaw the food before cooking. Cooking meat from the frozen state seals in the juices.

I always coat lamb cutlets and veal steak for wiener schnitzel with egg and breadcrumbs before freezing. Place a layer of plastic between each coated cutlet or veal steak, then pack in a plastic bag, remove air, seal and freeze.

POULTRY should always be plucked, cleaned and chilled in a refrigerator for at least 16-18 hours before packaging for freezing. Wash giblets separately. Insert inside the dressed bird. Wrap the entire bird in strong plastic or place in a polythene bag, remove the air and seal and freeze. To freeze pieces of poultry, bring the bird to the boil first, to make it easier to cut neatly. Cut into pieces and package as for steak and chops. Portions of poultry may also be dipped in egg and breadcrumbs before freezing. Place on a flat smooth tray, freeze and package in a plastic bag in the usual way. Thaw poultry portions for at least 2 hours at room temperature before cooking. Whole birds are best thawed completely before cooking.

GAME should be bled, cleaned and chilled as soon as possible after the kill. Always hang large carcases 8-10 days before cutting up like beef. Freeze smaller game whole or in portions like poultry (see above).

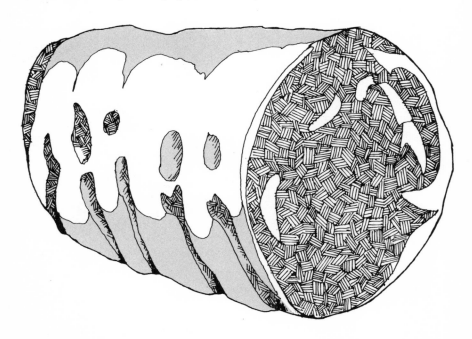

FREEZING FISH

Try to freeze fish as quickly as possible after catching, as it deteriorates rapidly. Before freezing, prepare fish as for cooking. Fish can be frozen whole, filleted or cut into steaks.

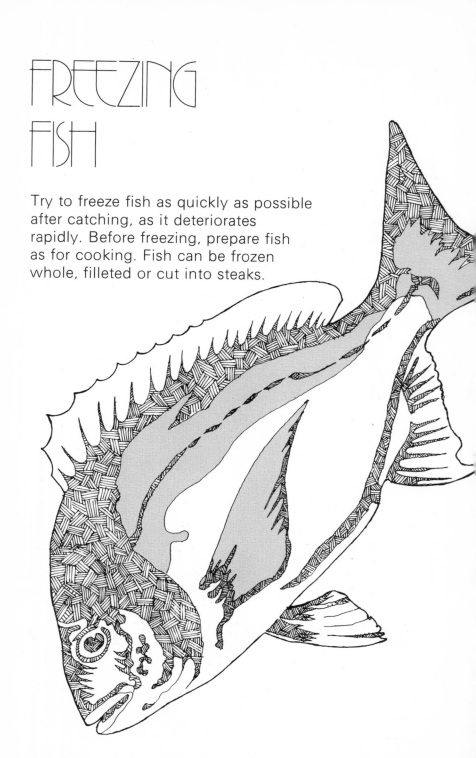

WHOLE FISH should be gutted and cleaned thoroughly. Some authorities recommend freezing whole fish with the scales on for a better flavour. FILLETS may be dipped into beaten egg and coated with breadcrumbs if desired, (the frozen fillets of fish are then ready to be cooked later without thawing). Lay individual FISH FILLETS AND STEAKS and whole cleaned fish onto a smooth tray and freeze for approximately 2 hours. Once frozen, remove from tray and pack in a good quality plastic or polythene bag. Remove the air from the bag before sealing, as this will help to prevent the fish from drying out and excess ice forming inside the packet. DO NOT THAW—always cook fish from the frozen state.

Cod, tuna, New Zealand groper and Australian mullet are FATTY OR OILY FISH. Dip this type of fish into lemon juice and water before freezing. Use 1 large lemon to 1 pint of cold water or $1\frac{1}{4}$ teaspoons of ascorbic acid powder to 2 pints of cold water. Dipping oily fish before freezing helps to retain the natural flavour.

SHELL FISH, cooked or uncooked, should only be frozen if completely fresh. When opening shell fish, prevent the grit from falling into the contents by washing the shells thoroughly in cold water before opening. To freeze individual oysters, mussels, pipis, toheroas etc., simply force the shells open, letting the liquid and fish fall into ice cube trays or similar divided containers. Add a few drops of lemon juice to each fish to help seal the colour. Once frozen, tip cubes into a strong plastic bag, remove the air and seal well. Shell fish will keep in good condition for up to 6 months. Thaw in a refrigerator and serve chilled raw or cook in the usual way.

Crayfish (and lobster)—for special dishes it is desirable to keep the tails whole. Remove cold cooked tails from shells. Dip each tail into a brine solution (2 oz salt to 2 pints water). Spread out each wet tail separately onto a smooth tray. Freeze for 1 hour, then dip in solution again and freeze again. Repeat dipping and freezing process after 20 minutes. Wrap each frozen tail separately in waxed paper. Package as many as you like into a large plastic bag. Remove the air and seal well. If tails are to be

recooked later, do not thaw, cook from the frozen state, allowing approximately 15-20 minutes extra cooking time. Tails prepared and packaged like this can be kept for up to 3 months. If kept longer they will toughen and dry out.

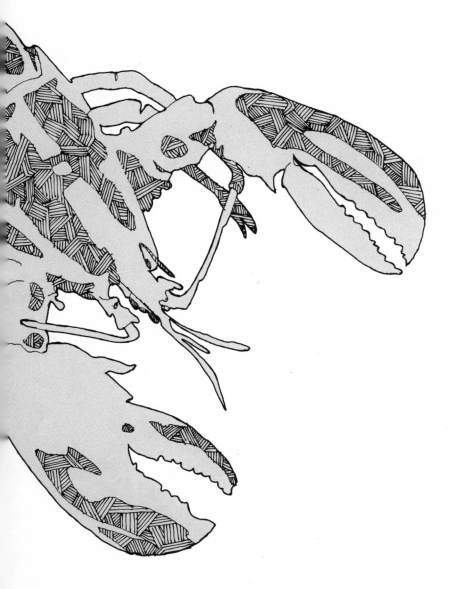

FREEZING VEGETABLES

Providing vegetables are treated properly they will freeze particularly well, retaining more flavour and goodness than by preserving in any other method. Learn the four basic rules dealt with in this section and you cannot fail. BLANCHING, CHILLING, PACKAGING and correct COOKING. Vegetables cannot be frozen raw. They must be blanched first, whilst fresh and tender. Blanching vegetables helps to seal in the colour, flavour and nutritive value. Raw vegetables stored in a freezer will quickly lose quality.

(1) BLANCHING

This is quite easy. Put approximately 1 lb of fresh clean vegetables into a blanching basket sitting in a large quantity of rapidly boiling water. Do not add salt. Place a tight fitting lid in position and wait until the water returns to the boil before starting the blanching time. Blanching times vary depending on the size of the individual pieces of vegetable—$1\frac{1}{2}$ minutes is ample blanching time for PEAS, sliced BEANS or CARROT rings— $1\frac{1}{2}$ minutes is also ample for shredded CABBAGE, SPINACH and SILVERBEET. When blanching larger pieces of vegetables simply lengthen the blanching time e.g. 1-inch pieces of green beans would take 2 minutes, whole beans or small thin carrots would take 3 minutes. CORN ON THE COB, being larger still, would take 5-8 minutes depending on the size. Consequently judging the blanching time required becomes easy once the size factor is remembered. For example a CAULIFLOWER left whole, would need blanching for 6-9 minutes. However, if the cauliflower is broken into flowerettes 2-3 minutes would be ample. POTATOES, KUMARA AND PUMPKIN should be half cooked before freezing. Blanching is insufficient for this type of vegetable.

Chill quickly then freeze on trays for 2 hours before packaging and returning to the freezer. BEETROOT can be successfully frozen by half cooking without peeling. Chill quickly, then peel and slice or dice before packaging in meal size quantities. I suggest the flat pack method (see page 107). No further cooking is required, as the deep freeze process has a tenderising effect. Thaw beetroot when required and serve as desired. MUSHROOMS also freeze well. Fry lightly in butter and spread out to cool on cold plates. Deep freeze cold mushrooms by the flat pack method, (see page 107). As an alternative method, simmer mushrooms in milk until tender. Thicken with rice flour for best results, leave to cool completely before packaging in meal-size quantities. To serve mushrooms, simply reheat whilst still frozen and use in your favourite recipe. Frozen cooked mushrooms will keep in good condition for up to 4 months.

(2) CHILLING
When blanching time is complete, lift blanching basket from boiling water and gently transfer vegetables into a colander sitting in ice cold water. A block of ice placed in a sink full of cold water will speed up the chilling process. Vegetables should not take any longer to chill than they do to blanch, if you use ice in the chilling

water. Freeze 2 pint unbreakable
bowls of cold water to make blocks
of ice.

(3) PACKAGING
Once vegetables are chilled, remove
from water and drain. For best results
pat dry with a clean tea towel. I
recommend the free flow method of
packaging or the flat pack
(see pages 106 and 107).

(4) COOKING
So many frozen vegetables are spoilt
by overcooking. Remember, they have
already been tenderised twice. Once
when blanched, and again whilst
stored below zero in your freezer.
Therefore, very little cooking is
required. DO NOT THAW—simply
drop frozen vegetables into boiling
salted water. When water returns to
the boil, lower the heat then simmer
for one third or less of the cooking
time normally given to fresh vegetables.
Although all frozen vegetables should
be cooked from the frozen state I
find corn on the cob the exception
to the rule, because it takes such a
long time to defrost in the centre.
I recommend thawing corn cobs
for at least 2 hours before cooking
in boiling salted water for 5-6
minutes. Potatoes, kumara and
pumpkin may be placed into hot fat
whilst still frozen and roasted (baked)
in a hot oven, or cooked in a deep
fryer until golden brown.

FREEZING FRUIT

Preserving fruit by deep freezing is easy and very convenient because fruit can be frozen raw and eaten raw all the year round. Raw fruits which are low in acid content need treating before freezing to prevent discoloration on thawing. To help you identify high or low acid type fruits I have prepared the chart below:

LOW ACID FRUITS

Raw fruits low in natural acid tend to discolour and go brown once cut. To prevent discoloration I recommend adding ascorbic acid, because it is nature's vitamin C and so we add a vitamin to the fruit.

APPLES
APRICOTS
AVOCADOS
BANANAS
CHERRIES
(SWEET)
FEIJOAS
NECTARINES
PEACHES
PLUMS
(WHITE
FLESH)

The low acid fruits listed here should be cored, stoned and peeled (where applicable). Place fruit into acid solution, 1½ teaspoons of ascorbic acid powder to 2 pints of ice cold water. Soak for approximately 5 minutes. Drain and use either the RAW FRUIT

PEARS	SUGAR METHOD or
PERSIMMONS	the RAW FRUIT
QUINCES	SYRUP METHOD (see page 36) for freezing.

HIGH ACID FRUITS
Raw fruits high in acid content freeze particularly well, and do not require soaking in ascorbic acid before freezing, as they contain enough of their own acid to prevent discoloration on thawing.

BLACKBERRIES
BOYSENBERRIES
CHERRIES
(SOUR)
RASPBERRIES
STRAWBERRIES
RHUBARB
TAMARRILLOS
GRAPEFRUIT
ORANGES
CHINESE
GOOSEBERRIES
GUAVAS
GOOSEBERRIES
MELON
PAPAYA
(PAWPAW)
PINEAPPLE
PLUMS (DARK
FLESH)
RED AND BLACK
CURRANTS
PASSIONFRUIT
BANANA and
PASSIONFRUIT
LEMON JUICE

The high acid fruits listed here can be frozen using either the DRY RAW FRUIT METHOD, the RAW FRUIT SUGAR METHOD or the RAW FRUIT SYRUP METHOD (see page 36). Prepare fruit as for eating i.e. peel, core, or stone (where applicable). Wash berries in ice cold water, drain and dry in a clean tea towel before freezing in one of the above mentioned methods. Red and black currants are excellent frozen in the DRY RAW FRUIT METHOD ready for making jam or pies all the year round.

DRY RAW FRUIT METHOD
Wash fresh ripe fruit in ice cold water, drain and dry carefully. Prepare as for eating. Peel, core, or stone (where applicable). Place meal or pie-size quantities into good quality plastic

or polythene bags, pat gently into a
flat pack shape. Remove the air with
a vacuum pump, seal with a wire
twist. Alternatively, use the free flow
method of packaging (see page 106).
To thaw, simply thaw in a
refrigerator and serve chilled (see
pages 110 and 111) or use without
thawing in all types of cooking.

RAW FRUIT SUGAR METHOD
Wash fresh ripe fruit in ice cold water,
drain and dry carefully. Prepare as for
eating. Peel, core or stone (where
applicable). Use a good sized bowl
and 2-3 oz of castor sugar to 1 lb of
fruit. Prepare 1 lb or 2 lb lots of fruit
at one time to prevent unnecessary
crushing. Place fruit into bowl and
sprinkle with sugar, shake bowl gently
from side to side to distribute the
sugar. Guard against adding too much
sugar as an excess will not freeze.
Should extra sweetening be required
it can be added after thawing. I
recommend the flat pack method of
packaging sugared fruits in meal-sized
quantities or use any of the special
plastic or wax coated containers with
tightly fitting lids (see pages 104 and
105). Seal, label and freeze. To thaw,
simply thaw in a refrigerator for 9-10
hours and serve chilled (see pages
110 and 111).

RAW FRUIT SYRUP METHOD
Prepare a sugar syrup, approximately

2 cups water to 1 cup sugar, and leave to cool completely. Dissolve 1¼ teaspoons ascorbic acid powder in every 2 pints of cold syrup. Now wash fresh ripe fruit in ice cold water, drain and dry well. Prepare as for eating. Peel, core or stone (where applicable). Add fruit to syrup, leave to stand for 5 minutes then package in meal-size quantities. Plastic or waxed containers with tightly fitting lids are ideal (see pages 104 and 105). Remember to leave approximately 1-inch head room for the liquid to expand. Seal carefully, label and freeze. To thaw, simply thaw in unopened container in a refrigerator, serve chilled (see pages 110 and 111).

COOKED FRUIT
Cooked fruit can also be frozen as a means of preserving. Caution on three points:
(1) Do not overcook fruit to be frozen. Remember a freezer has a tenderising effect on all foods. You will find that half the normal cooking time is ample for fruit which is to be frozen.
(2) Wait until fruit is completely cold before packaging and freezing.
(3) Remember to leave approximately 1-inch head room in container for expansion.
Freeze meal or pie-size quantities in plastic or waxed containers with tight fitting lids. To thaw (see pages 110 and 111).

FREEZING COOKED FOOD

A selection of meals stored in your deep freeze can be such a help, especially when catering for the unexpected guest or on occasions when you may have to go away, leaving the family to look after themselves. Remember to label all such meals clearly with the date, contents and reheating instructions.

Meals such as casseroles and stews, etc. should be partially thawed for at least 1 hour before reheating in a moderate oven (350 °F. gas No. 4). The time allowed for reheating will, of course, vary depending on the quantity. A 'meal-size' for 6 usually takes 1 hour if partially thawed first.

Freeze pies in aluminium foil pie plates for convenience. They may be frozen baked or unbaked. Cooked pies naturally take less time to heat through. Once again, preheat your oven (400 °F. gas No. 5) with a baking tray in place. Place the frozen pie onto a hot baking tray. Bake for 35-40 minutes. The direct heat will seal the pastry base before the filling thaws out, so preventing a wet base.

Unbaked pies may be baked unthawed
according to the recipe allowing
15-20 minutes extra.

Soup may be frozen easily. Freeze in
meal-size portions. To serve, heat
slowly without thawing.

IMPORTANT POINTS

(1) Always use rice flour to thicken gravy, sauces, stews, casseroles, etc., as ordinary flour or cornflour tends to separate on thawing. Gravy should not be stored for too long because of its high fat content.

(2) Remember your deep freeze has a tenderising effect on all food, therefore only partly cook any vegetables to be included in dishes to be frozen.

(3) White potatoes do not freeze well, so omit from frozen stews and casseroles.

(4) Development of poor flavours is often caused by flavourings such as onion (in large quantities), herbs, spices and wine, so it is wiser to add these after thawing.

(5) Always make sure meals are completely cold before packaging to freeze, as warm meals placed into a freezer cause condensation and loss of quality.

Whenever possible use tinfoil or aluminium foil dishes for storing pre-cooked meals in your freezer, as these dishes can be used to reheat meals in a hurry without thawing i.e. straight from freezer to oven, and add 20 minutes to reheating time. Alternatively, there is a type of 'temperature-resistant' casserole on the market which may be taken straight from the deep freeze and placed immediately in a hot oven.

A WESTINGHOUSE FROST FREE
REFRIGERATOR FREEZER

A WESTINGHOUSE
CHEST FREEZER

A WESTINGHOUSE FROST FREE
UPRIGHT FREEZER

A KELVINATOR NO FROST
REFRIGERATOR FREEZER ·

FREEZING BAKED FOOD

Most people associate a deep freeze with meat, fish, vegetables and fruit. However, baked foods keep equally well below zero. Make full use of your deep freeze this year by preparing cakes, biscuits, pies, savouries and even bread, ready for the busy Christmas period or parties. If you are expecting visitors, think of the convenience of having morning and afternoon teas ready in advance. No fuss or bother, simply take items from your freezer to thaw, whilst you sit relaxed, and enjoy your visitors' company.

CAKES

Package unfilled, un-iced cakes in plastic bags, seal well, label and freeze. It is advisable to protect them with a double layer of plastic. Cake batter may be frozen in greased cake tins. Wrap and seal well and freeze. To bake, remove packaging and place frozen cake in oven until cooked.

CREAM FILLED CAKES

Both fresh cream and synthetic cream may be deep frozen successfully. If fresh cream is used, I suggest leaving it to stand in a refrigerator, after whipping, for at least 2-3 hours, to allow the water content to settle at the bottom of the bowl, then use to fill cakes and sponges. This method prevents a wet base on thawing. For best method of packaging see page 104.

ICED CAKES

Iced cakes should be frozen first, then wrapped and sealed for storing in the freezer. Water type icing tends to become very brittle after freezing. Synthetic icing becomes very sticky. I suggest keeping to a simple butter icing using any type of fruit juice (lemon, orange, grapefruit, passionfruit, etc.).

SIMPLE BUTTER ICING

1 cup icing sugar
1 tablespoon softened butter
1 tablespoon fruit juice

Sift icing sugar into a mixing bowl. Add softened butter and fruit juice and mix with a wooden spoon until smooth and light in texture.

BISCUITS

If the basic foundation recipes are used, biscuits can be frozen successfully, either baked or as unbaked dough. If baked, freeze by the free flow method (see page 106), package carefully, seal well, label and store in freezer. Unbaked dough may be shaped into round rolls or a cylinder shape, then wrapped and frozen. To bake, slice frozen dough thinly, as required, place on a baking tray and bake according to recipe.

SAVOURIES

All types of delicious savouries can be
frozen either baked or unbaked.
Let the type of filling be your guide
as to how long to keep them (see
pages 100, 101 and 109). Freeze on
trays for approximately 2 hours before
packaging, to prevent crushing when
stored in your freezer. Frozen savouries
can be glazed with a little beaten egg
then placed into a hot oven (400 °F
gas No. 6), straight from the freezer,
for approximately 20 minutes.

BREAD

All types of bread and rolls freeze well,
in fact several of my friends insist
that bread improves with the freezing.
However, good packaging materials
must always be used (see pages
104 and 105). Poor packaging will
result in a drying out and change in
flavour. Frozen sliced bread may be
separated easily whilst still frozen,
which is a great advantage when
small quantities are required. If a
whole loaf of bread is required in a
hurry, simply preheat oven and bake
for approximately 20 minutes. If not
in a hurry leave bread in wrapping to
thaw at room temperature.

Unbaked bread dough may be frozen.
Wrap and seal well before freezing
in the tin to be used for baking. Thaw
before baking.

FREEZING DAIRY FOOD

All dairy foods attract the flavours of other foods, and so develop an unpleasant taste. Therefore, good packaging is essential to protect the delicate flavours of dairy produce. Remember to label clearly with date, contents and quantity.

BUTTER, MARGARINE AND LARD
These will keep below zero in a good condition for up to 6 months, if properly packaged. I recommend a double wrapping for extra protection.

CHEESE
Cheese has a tendency to crumble when thawed. However, all types of cheese will keep their flavour and goodness for up to 6 months in the freezer. Wrap in waxed paper or foil, then place inside a good quality plastic bag, remove the air and seal well.

ICE CREAM
This also needs good packaging to prevent 'off flavours'. I suggest home-made ice cream is placed into a plastic container with a tight fitting lid. Remove commercial ice cream from cardboard container to prevent a possible waxy flavour and package as described for home-made ice

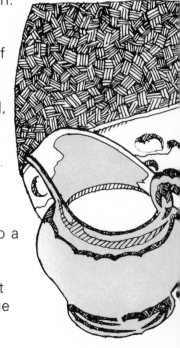

cream. Some commercial ice cream is sold already packaged in a plastic container.

EGGS

YOLKS: To 1 cup of yolk, mix in 1 teaspoon of salt, or 1 tablespoon of sugar. Freeze in a plastic or polythene container or plastic bags standing in a box as a mould. Remember to leave head space for expansion.

WHITES: These keep in your freezer particularly well, salt or sugar is not necessary. Place egg whites into a waxed or plastic container, leave 1-inch space for expansion, seal and label with details.

WHOLE EGGS: Gently mix whites and yolks together—do not whip. Add 1 teaspoon of salt to each 2 cups of eggs. Ideal for omelets or scrambled eggs. As an alternative add 1 tablespoon of sugar, and use for cakes and custards later.

1 tablespoon of egg yolk	= 1 egg yolk
2 tablespoons of egg white	= 1 egg white
3 tablespoons of whole egg mix	= 1 fresh egg
1 cup of mixed whole egg	= 5 fresh eggs

FREEZING FOR CHRISTMAS

The Christmas period is a happy but busy time for all of us. Make full use of your deep freeze in preparing all sorts of time saving items.

(1) Prepare large quantities of ICE ready for serving drinks to visitors.

(2) Several types of SAVOURY TOPPINGS for biscuits, plus a variety of cream cheese dips can be prepared, packaged and kept in your freezer for up to 4 weeks. Prepare and freeze quantities of grated cheese, chopped ham and CROÛTES CHEESE STRAWS will keep crisp and fresh in your freezer. CANAPÉS such as salmon or seafood on crisp croûtes, also freeze well. DO NOT freeze olives, capers or salad vegetables.

(3) All types of SOUPS may be prepared, packaged and frozen. Remember to use rice flour when thickening is required.

(4) Remember the TURKEY can be purchased weeks or even months before Christmas. Properly packaged and deep frozen, it will keep in perfect condition for up to 12 months. CAUTION: turkey and chicken are best thawed out before cooking. A turkey will take 2-3 days to thaw at room temperature, depending on size. DO NOT remove the protective wrapping during the defrosting period. Cook as usual. Leftover cold turkey can be packaged and returned to your freezer ready to make all sorts of delicious dishes at a later date. Storage time for cooked turkey is 4 months.

(5) A vast selection of VEGETABLES may be prepared and stored in your deep freeze, (see page 28).

(6) Traditional CHRISTMAS PUDDING, fruit MINCE PIES and even the CHRISTMAS CAKE can be prepared months in advance, packaged and deep frozen. Even the BRANDY SAUCE can be prepared in advance to save time on that busy day. Remember to use rice flour to thicken, as a cornflour sauce tends to separate after freezing.

(7) Your CHRISTMAS HAM can be purchased, when reasonably priced, packaged and frozen. However, remember the fat on ham has a tendency to go rancid if kept longer than 6 months. Thaw for at least 3 days without removing the protective wrapping. Cook as usual. Leftover cooked ham can also be packaged and re-frozen for use later. Storage time for cooked ham is 4 months.

(8) All types of NUTS can be frozen, if you use a good packaging material. Storage time is approximately 6 months.

CHEESE STRAWS

Makes: 18
Cooking time: 10 minutes
Temperature: 375-400 °F. gas No. 5-6

4 oz (1 cup) plain flour
pinch of dry mustard
pinch of salt
pinch of cayenne pepper
3 oz butter
3 oz Parmesan cheese, grated
1 egg yolk mixed with 2 teaspoons cold water

Sift flour and seasonings. Cream butter in a mixing bowl until soft and white. Add flour, grated cheese and enough combined egg and water to mix to a stiff dough.

Roll out pastry thinly, cut into straws approximately 4-inches long by $\frac{1}{8}$-inch wide. Place on greased baking trays and bake in a moderately hot oven for 10 minutes or until golden brown. Cool on a wire cooling tray.

To freeze: Wrap carefully in thin film of plastic, place in an airtight plastic container or plastic bag (remove air), seal well, label and freeze.

To serve: Thaw Cheese Straws in wrappings at room temperature for about 1 hour.

ICED AVOCADO SOUP

Serves: 4

2 large ripe avocado pears
1½ pints chicken stock or water and chicken stock cubes
pinch of salt, pepper and ground nutmeg
¼ pint cream
extra cream for serving

Peel avocado pears, remove stones and mash.
Place in a saucepan with chicken stock, heat gently.
Pass through a sieve or blender to remove any
lumps. Return to saucepan, add seasonings and
bring slowly to simmering point. Stir in cream,
remove from heat and cool.

To freeze: Pour into a plastic container allowing
½-inch headspace. Seal well, label and freeze.

To serve: Thaw at room temperature for 1 hour.
Pour into bowls and top each serving with a little
extra cream.

CREAM OF TURKEY SOUP

Serves: 8

2 pints turkey stock (see below)
2 oz butter
2 oz (4 tablespoons) plain flour
1 pint scalded milk, (or ½ milk and ½ cream)
1 stalk celery, finely chopped
salt and pepper

Make stock by placing turkey carcase in a large saucepan and covering with cold water. Bring slowly to the boil, cover and simmer gently for approximately 2 hours. Cool, strain and reserve stock.

Melt butter in a large saucepan, stir in flour, cook for approximately 1 minute. Add 1 pint of stock and stirring continuously, bring to the boil. Add remaining stock, milk, celery and salt and pepper to taste. Bring to the boil and simmer gently for 5 minutes, cool.

To freeze: When cold, pour into a plastic container, allowing ½-inch headspace. Seal well, label and freeze.

To serve: Thaw at room temperature for 2 hours, heat and serve immediately.

HAM & PEA SOUP

Serves: 6-8

8 oz split peas
1 oz butter
2 oz ham, diced (cut from bone)
2 onions, sliced
1 large carrot, diced
4 pints water
1 cup sliced celery
1 bay leaf
1 ham bone
2 teaspoons salt

Soak peas overnight in cold water to cover, drain. Heat butter in a large saucepan, sauté ham, onions and carrot until onion is golden. Add remaining ingredients and bring to the boil. Simmer, covered, for $2\frac{1}{2}$-3 hours. Remove bay leaf and ham bone from soup, adjust seasoning if necessary. Cool.

To freeze: When cold, pour soup into plastic container, allowing $\frac{1}{2}$-inch headspace. Seal well, label and freeze.

To serve: Thaw at room temperature for 1 hour. Heat gently and serve when hot.

SAUSAGE MEAT STUFFING

liver from bird to be stuffed
8 oz sausage meat
2 oz soft white breadcrumbs
salt and pepper
pinch of grated nutmeg
stock
$\frac{1}{2}$ teaspoon finely chopped fresh herbs or
$\frac{1}{4}$ teaspoon dried herbs

Finely chop liver. Combine all ingredients, excluding herbs, adding sufficient stock to bind stuffing together.

To freeze: Place in an airtight plastic container or plastic bag (remove air), seal well, label and freeze.

To serve: Thaw in refrigerator overnight. When ready to use, add herbs and mix in thoroughly.

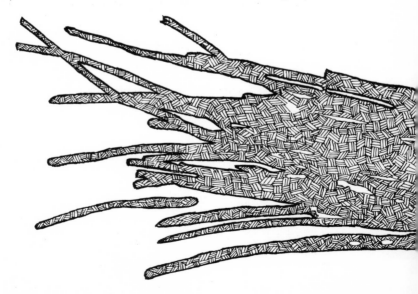

CHESNUT STUFFING

2 lb chestnuts
$\frac{1}{4}$-$\frac{1}{2}$ pint stock
2 oz butter
salt and pepper
pinch of ground cinnamon
$\frac{1}{2}$ teaspoon sugar

Slit chestnuts and bake or boil for 20 minutes, remove shells and skins. Place chestnuts in a saucepan and just cover with stock, simmer until tender. Put through a sieve or blender. Add remaining ingredients and mix in sufficient stock to make a soft dough.

To freeze: Place stuffing in an airtight plastic container or plastic bag (remove air), seal well, label and freeze.

To serve: Thaw in refrigerator overnight. Stuff turkey.

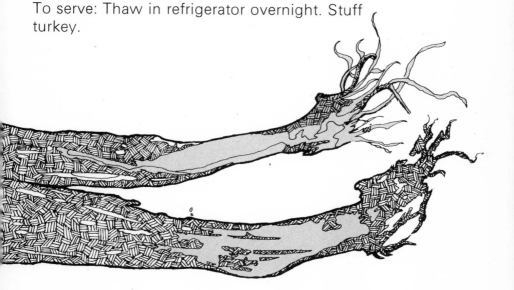

BREAD SAUCE

1 large onion
2 cloves
1 bay leaf
4 peppercorns
$\frac{1}{2}$ pint milk
2 oz soft white breadcrumbs
$\frac{1}{2}$ oz butter
salt and pepper

Place onion, spices and milk in a saucepan. Bring slowly to the boil, remove from heat, cover and allow to stand for approximately 30 minutes. Strain liquid. Add breadcrumbs, butter and salt and pepper to taste, cool.

To freeze: Place sauce in an airtight plastic container allowing $\frac{1}{2}$-inch headspace. Seal well, label and freeze.

To serve: Thaw in container at room temperature for 1 hour. Place in saucepan and reheat gently.

PLUM PUDDING

Serves: 6

4 oz (1 cup) plain flour
pinch of salt
1 teaspoon baking powder
4 oz soft white breadcrumbs
4 oz suet, finely chopped
4 oz ($\frac{2}{3}$ cup) brown sugar
3 oz raisins
4 oz currants
$\frac{1}{4}$ teaspoon grated nutmeg
1 egg
$\frac{1}{4}$ pint milk

Grease a 2-pint pudding basin. Sift flour, salt and
baking powder into a mixing bowl. Add breadcrumbs,
suet, sugar, raisins, currants and nutmeg. Mix
thoroughly, adding beaten egg and milk. Place
mixture in basin, cover securely with aluminium
foil. Boil for 4-5 hours.

To freeze: Turn pudding out of basin while hot.
When cool, wrap in a thin sheet of clear plastic
and place in a strong plastic bag. Remove air from
bag with vacuum pump. Seal carefully, label and
freeze.

To serve: Thaw pudding in refrigerator overnight.
Place in pudding basin, cover and reboil for 3-4
hours before serving.

RUM SAUCE

¼ pint cream
2 egg yolks
1 tablespoon brown sugar
2-3 tablespoons rum

Place all ingredients in a mixing bowl. Place over a saucepan of boiling water and whisk constantly until mixture thickens.

To freeze: Pour into a plastic container allowing ½-inch headspace. Seal well, label and freeze.

To serve: Thaw at room temperature for 1 hour. Place in a bowl and reheat gently over a saucepan of boiling water.

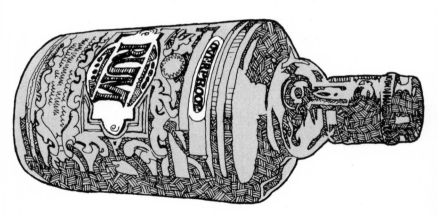

MINCE PIES

Makes: 36
Cooking time: 15-20 minutes
Temperature: 400-450 °F. gas No. 6-7

½ quantity short crust pastry (see page 66)
Filling:
2 large cooking apples
2 oz sultanas
2 oz currants
1 oz mixed peel
2 oz (⅓ cup) brown sugar
½ oz butter, melted
pinch of nutmeg
grated rind and juice of 1 lemon
egg white or water and sugar for glazing

Peel, core and grate apples. Combine all ingredients in a large mixing bowl, mix together thoroughly.

Roll out pastry and cut into rounds with a biscuit cutter. Place half in greased patty tins, fill with mincemeat, and cover with remaining rounds of pastry. Glaze pies with beaten egg white or water and sprinkle with sugar.

Bake in a hot oven for 15-20 minutes or until pies are golden brown on top. Remove from patty tins and cool on a wire cooling tray.

To freeze: Place pies in a single layer in a flat shallow plastic container, seal well, label and freeze.

To serve: Thaw pies at room temperature for 2 hours. Place on a baking tray and heat in a moderate oven (350-375 °F. gas No. 4) for 10-15 minutes.

FREEZING FOR A LONG WEEKEND

Catering for a long weekend, such as the Easter weekend or a bank holiday weekend, is no small undertaking. The hard way is to leave all the shopping to the last minute and to spend the weekend cooking. The easy way, with a little thought and planning on your part, is to prepare some dishes well in advance and store them in the freezer. Many suitable dishes may be prepared in advance and frozen, all ready to assemble and serve quickly without time and fuss.

Prepare and freeze a selection of sandwiches (not egg or salad), pies and cakes, ready for a picnic or to serve to unexpected visitors for morning or afternoon tea. Prepare the baby's food in advance to save time. As a special treat for the children, freeze wedges of melon or slices of orange, pierced with toothpicks, ready to hand out like ice blocks. Make sure you have plenty of fresh bread stored in your freezer, also a supply of steaks and chops ready for the barbecue. Prepare, cook and freeze savoury dishes, vegetables, fruit desserts and ice cream for the main meals of the weekend.

Imagine that you are having your favourite friends to stay as house guests for a long weekend—a married couple with two children, just like you. Try the following menu ideas and you will find that you too can have a holiday and enjoy the company of your guests at leisure. Make full use of your deep freeze and have a holiday yourself on the next long weekend.

Friday evening	French Onion Soup
	Spaghetti with Bolognese Sauce
	Tropical Surprise
	Coffee
Saturday lunch	Vichyssoise
	Cornish Pasties
	Tossed Salad
	French Bread Sticks or Bread Rolls
	Fresh Fruit
	Fruit Drinks and Milk
Saturday dinner	Watermelon Cocktail
	Chicken in Wine
	New Potatoes
	Green Peas
	Pawpaw Sherbet
	Cheese
	Coffee
Children	Grilled Lamb Chops
	Potatoes
	Carrot rings
	Peas
Sunday lunch	Veal with Ham and Cheese
	Creamed Potatoes
	Tomatoes
	Zucchini
	Strawberries and Ice Cream
	Coffee

FRENCH ONION SOUP

Serves: 8
Cooking time: 20 minutes

2 lb white onions
2 oz butter
1 oz (2 tablespoons) flour
3 pints beef stock (made from stock cubes)
salt
freshly ground black pepper
slices of French bread
8 oz Cheddar or Swiss cheese, grated

Finely slice onions and sauté them in butter in a large heavy saucepan, until golden brown. Stir in flour. Gradually add stock, cover and bring to the boil. Season with salt and pepper. Simmer for 15 minutes. Cool rapidly.

To freeze: Remove any fat from top of soup. Pour into a seal top plastic container, allowing 1-inch headspace. Seal, label and freeze.

To serve: Remove soup from container and thaw in a saucepan over a very slow heat. Bring to boiling point and add more seasonings if necessary. Pour into individual heatproof tureens or soup bowls. Spread slices of French bread with a little butter and cover with grated cheese. Place one slice in each tureen. Place tureens in a hot oven (400-450°F. gas No. 6-7) for 5-10 minutes to melt cheese. Serve at once.

BOLOGNESE SAUCE

Serves: 8
Cooking time: 1¼ hours

8 oz white onions
2 tablespoons oil
2 lb lean chuck steak, minced
4 tablespoons tomato purée
1 × 15 oz can peeled tomatoes
1 teaspoon salt
freshly ground black pepper
½ tablespoon brown sugar
½ teaspoon dried mixed herbs
4 fl oz (½ cup) red wine
2 cloves garlic
1 bay leaf
grated Parmesan cheese for serving

Chop onions finely and fry in oil in a large heavy saucepan, until golden brown. Add meat and continue frying and stirring until meat is cooked. Add the tomato purée, peeled tomatoes, seasonings, herbs and wine. Simmer, covered, for 1 hour, stirring occasionally. Cool rapidly and remove any excess fat.

To freeze: Place in plastic containers, allowing ½-inch headspace. Seal, label and freeze.

To serve: Remove sauce from container in a block and thaw in a saucepan over hot water. Crush or chop garlic and add to sauce. Adjust seasonings to taste and add bay leaf. A little extra red wine may be added at this stage. When thawed completely, cook for an additional 30 minutes.

To serve spaghetti with Bolognese Sauce: Cook spaghetti when required, drain well, serve and pour sauce over, topping the sauce with grated Parmesan cheese. Serve piping hot.

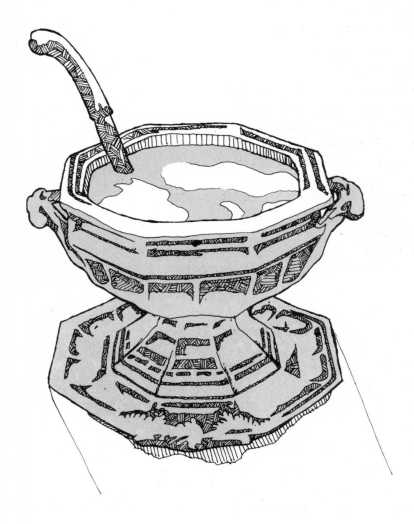

TROPICAL SURPRISE

Serves: 4

$\frac{1}{2}$ pint rosé wine
2 cups chopped fresh fruit (apples, pears,
bananas, pineapple, pawpaw, strawberries)

Pour wine into a large bowl. Prepare a variety of
fresh fruit accordingly—peel, core, remove seeds
—and cut neatly into small pieces. Place pieces of
fruit into the wine as soon as cut.

To freeze: Package fruit and wine into a plastic
container, allowing 1-inch headspace. Seal well,
label and freeze.

To serve: Thaw in refrigerator and serve with fresh
whipped cream sprinkled with chopped nuts.

VICHYSSOISE

Serves: 8
Cooking time: 1 hour

4 oz butter
6 leeks
2 white onions
3 pints chicken stock (made from stock cubes)
1¼ lb potatoes
2 stalks celery
1 teaspoon salt
freshly ground pepper
1 pint cream
chopped chives

Melt butter in a large saucepan. Slice leeks and onions and sauté them in the butter. Add the stock and peeled, diced potatoes, chopped celery, salt and pepper. Cover and cook until tender, approximately 45 minutes.

Rub the cooled soup through a sieve or mix to a purée in a vitamiser or blender. Cool rapidly and remove any fat which rises to the surface.

To freeze: Place soup in plastic containers, allowing 1-inch headspace. Seal, label and freeze.

To serve: Thaw the soup in the container allowing about 12 hours in the refrigerator. Keep chilled before serving and, at the last minute, blend in the cream. Serve in mugs topped with a spoonful of chopped chives.

CORNISH PASTIES

Makes: 12

Cooking time: 40-45 minutes
Temperature: 400-450°F. gas No. 6-7 reducing
to 375-400°F. gas No. 5-6

1 quantity of short crust pastry
beaten egg yolk for glazing
Filling:
1 lb lean chuck steak, minced
2 white onions, chopped
2 potatoes, peeled and diced
1 small carrot, diced
$\frac{1}{2}$ small turnip, diced
2 teaspoons salt
freshly ground black pepper
2 tablespoons water
2 teaspoons chopped parsley

Place meat, vegetables, salt and pepper, water,
and parsley in a mixing bowl and mix well.

Divide pastry into 12 parts. Knead each lightly into
a ball and roll out to a round the size of a saucer.
Place a saucer upside down on the pastry and cut
out 12 even rounds. Place an equal portion of the
filling on each round. Brush around the edges of the
pastry with water. Join edges together over the top
of the mixture. Pinch a small, neat frill over the
join and shape into a crescent. Place pasties on a
baking tray and prick top with a fork. Glaze with
beaten egg yolk.

Bake in a hot oven for 10 minutes and then reduce

to moderately hot for 30-35 minutes. Place on a wire cooling tray to cool thoroughly.

To freeze: Place pasties in individual plastic bags and seal well before freezing.

To serve: Thaw pasties in plastic bags for 12 hours before serving. Serve cold with tomato sauce if desired and a tossed salad.

SHORT CRUST PASTRY

Makes: 1 lb

1 lb (4 cups) plain flour
2 teaspoons baking powder
1 teaspoon salt
8 oz butter
6-8 tablespoons water

Weigh and measure ingredients. Sift flour, baking powder and salt together into a mixing bowl. Add butter and rub it into the flour using the fingertips, until the mixture resembles the texture of breadcrumbs. Gradually add enough water to mix to a firm dough. Sprinkle board lightly with flour, turn dough on to board and knead lightly until smooth. Sprinkle flour on rolling pin. Roll pastry to size and shape required.

WATERMELON COCKTAIL

1 watermelon
8 tablespoons white wine or sherry
2 tablespoons brown sugar
$\frac{1}{2}$ teaspoon ground ginger

Choose a firm, ripe watermelon. Remove pips or seeds, and cut the melon flesh into cubes, wedges or balls, using a melon-ball scoop.

To freeze: Place pieces of melon on clean flat trays and place in freezer (free flow method, see page 106). When frozen, package in a plastic bag, remove air, seal well, label and return to freezer until required.

To serve: Thaw sufficient melon to serve 4 at room temperature for 1 hour. Place melon in individual cocktail glasses. Sprinkle with wine, sugar and ginger and serve chilled.

CHICKEN IN WINE

Serves: 4
Cooking time: 20 minutes

2 tablespoons oil
2 oz butter
8 chicken pieces (leg or breast)
2 rashers bacon, chopped
3 white onions, chopped
1 cup red wine
1 cup chicken stock (made from stock cube)
salt
freshly ground black pepper
8 oz button mushrooms
1 bay leaf
4 tablespoons chopped parsley

Heat oil and butter in a heavy based pan and fry chicken pieces until golden. Add bacon and onion and brown lightly. Add wine, chicken stock and salt and pepper to taste. Cover and simmer for 20 minutes. Cool rapidly and remove any surplus fat.

To freeze: Place in a seal-proof plastic container. Allow ½-inch headspace. Seal well, label and freeze.

To serve: Allow to thaw in container for 3-4 hours. Remove and place in a casserole together with the mushrooms, bay leaf, parsley, and a little extra red wine.
Gently heat in a slow oven (300-325 °F. gas No. 1-2) for 30 minutes. Continue cooking until chicken pieces are tender.

Remove chicken to a hot serving dish and thicken sauce with a roux, made by melting 1 oz butter and blending in 1 oz (2 tablespoons) flour. Gradually stir in chicken sauce and cook, stirring continuously, until sauce boils. Stir briskly for 2 minutes. Taste and adjust seasonings if necessary. Serve two pieces of chicken per person and pour sauce over. Serve hot with boiled new potatoes and minted peas.

PAWPAW SHERBET

Serves: 8

8 cups pawpaw pulp
4 cups castor sugar
juice of 4 large lemons
4 egg whites

Combine pawpaw pulp with sugar and lemon juice. Whisk 4 egg whites until they are stiff and fold them lightly into the pawpaw mixture.

To freeze: Pour mixture into ice trays and freeze without stirring.

To serve: Allow to thaw for 30 minutes before spooning into serving dishes. Decorate with strawberries from the freezer, and mint leaves. A dessert biscuit may be served with the sherbet.

VEAL WITH HAM & CHEESE

Serves: 8

8 thin veal steaks
4 oz (1 cup) plain flour
1 teaspoon salt
2 eggs
4 tablespoons milk
2-3 cups dried breadcrumbs
4 tablespoons oil
2 oz butter
8 slices ham
8 slices Swiss cheese

Flatten veal steaks with a cleaver or rolling pin. Sift flour with salt. Beat eggs with milk. Dip veal steaks in flour, then dip in egg mixture. Drain and toss each steak individually in breadcrumbs until evenly coated. Flatten with a knife and remove surplus crumbs.

Lightly fry veal steaks in a mixture of heated oil and butter, using more if required. Drain veal thoroughly and chill.

To freeze: Pack veal steaks with a thin film of plastic between each one. Pack in an airtight plastic container or plastic bag (remove air), seal well, label and freeze.

To serve: Remove veal from container or bag. Veal steaks should separate easily. Place on a foil lined baking tray and top each with a slice of ham, then a slice of cheese.

Bake in a moderate oven (350-375 °F. gas No. 4)
for 30 minutes or until veal heats through and
cheese is melted. Serve hot with mashed potatoes,
grilled tomatoes and zucchini rings.

STRAWBERRIES & ICE CREAM

Serves: 8

Take frozen strawberries from the freezer just 30
minutes before serving time, so that they are still
chilled, and pour a small amount of kirsch over
them.

Serve with ice cream. .

FREEZING FOR A CHRISTENING

Babies are wonderful, cuddly, adorable, but oh so time consuming! For the new mother, the very thought of the christening tea can be terrifying. Don't panic! Make full use of your deep freeze. If this is the first christening tea that you have given it may be a good idea to prepare everything in advance. Sit down and relax with paper and pencil and decide first of all on how many guests you will be inviting. Next, plan your menu, work out quantities and a shopping list. When a baby is tucked up for a long sleep you will have time to prepare at least one dish on your menu. Pop it into the freezer, well packaged of course, and store it in preparation for 'the big day'. The christening cake may also be stored in the freezer but this is not really necessary as rich fruit cakes keep very well indeed for many weeks, if stored correctly in an airtight container.

Try the following menu if you have to cope with a christening tea. You will have very little finishing off to do on the actual day if you follow the instructions carefully. The menu and recipes serve 12. Serve champagne for the toasts.

Pinwheel Sandwiches
Coffee Creams
Cream Puffs
Petit Fours

SEASONAL FRUITS WHICH MAY
BE FROZEN TO SERVE OUT OF SEASON

A CHRISTENING TEA
PACKAGED FOR THE FREEZER

THE CHRISTENING
TEA SERVED

A BIG BASKET THAT GLIDES OUT
OF A WESTINGHOUSE UPRIGHT FREEZER
PROVIDES QUICK ACCESS
TO PACKAGED MEAT

PINWHEEL SANDWICHES

Serves: 12

Pinwheel sandwiches:
12 slices white bread
12 slices brown bread
3 oz butter, melted
Cream cheese and gherkin filling:
4 oz packet cream cheese
4 tablespoons chopped gherkins
$\frac{1}{2}$ tablespoon finely chopped parsley
salt
freshly ground black pepper
Salmon filling:
1 x 6 oz can salmon
1 tablespoon mayonnaise
1 teaspoon lemon juice
salt
freshly ground black pepper

For sandwiches: Carefully remove all crusts from
bread and brush slices with melted butter.
For cream cheese filling: Soften cream cheese at
room temperature and blend all ingredients together.
Season to taste with salt and pepper.
For salmon filling: Drain salmon and remove all
bones. Finely flake salmon and combine well
with remaining ingredients.
To make sandwiches: Pile sandwich filling onto
buttered slices. Roll bread slices as for a Swiss roll,
making sure the bread is as tightly rolled as possible,
without losing the filling.

To freeze: Wrap the filled rolls of sandwiches in

foil. Place in airtight plastic bags, seal, label and freeze. Alternatively, freeze the foil wrapped sandwiches in sealed plastic containers.

To serve: Thaw sandwiches in wrappings in a refrigerator for about 2 hours. When thawed, unwrap and cut each sandwich into three parts. Place on plates and garnish with sprigs of parsley and wedges of tomatoes.

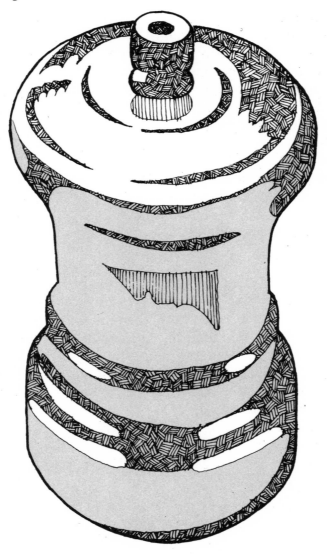

COFFEE CREAMS

Makes: 36
Cooking time: 12-15 minutes
Temperature: 325-350 °F. gas No. 3

Biscuits:
8 oz (2 cups) self-raising flour
$\frac{1}{4}$ teaspoon salt
1 egg
4 oz butter
3 oz ($\frac{1}{2}$ cup) castor sugar
1 teaspoon instant coffee
1 tablespoon hot water
Coffee butter cream:
1 oz butter
6 tablespoons sifted icing sugar
1 tablespoon full cream powdered milk
$\frac{1}{2}$ teaspoon instant coffee
3 teaspoons hot water

For biscuits: Sift flour and salt. Beat egg. Cream butter and sugar until light and fluffy. Beat in egg and add the instant coffee dissolved in hot water. Fold in flour and mix thoroughly. Force mixture through a forcing bag with a star pipe, or place teaspoons of mixture on greased baking trays, allowing room for spreading. Bake in a moderately slow oven for 12-15 minutes. Cool on a wire cooling tray and sandwich pairs together with coffee butter cream.

For butter cream: Cream butter, sugar and powdered milk. Add instant coffee dissolved in hot water. Mix thoroughly.

To freeze: Place biscuits on a tray in freezer and when frozen, lift off and carefully pack in polythene bags or containers. Seal, label and refreeze.

To serve: Thaw in containers for 30 minutes. Remove and place on serving dishes.

CREAM PUFFS

Serves: 12
Cooking time: 30-35 minutes
Temperature: 400-450 °F. gas No. 6-7 reducing
to 350-375 °F. gas No. 4

Choux pastry:
4 oz (1 cup) plain flour
1 cup water
pinch of salt
4 oz butter
3 eggs

For cream puffs: Sift flour. Place water, salt and
butter in a saucepan and bring to the boil. Remove
from heat. Add flour all at once and stir vigorously
until mixture leaves the sides of the pan. Allow to
cool slightly. Add 1 egg at a time, beating thoroughly
after each addition. The mixture is ready for baking
when the ingredients are well combined and the
mixture is shiny and smooth.

Using a forcing bag and a plain pipe, place
quantities of mixture on a greased baking tray,
leaving space for spreading. Bake in a hot oven for
20 minutes, then reduce temperature to moderately
hot for an additional 10 minutes. When risen,
golden brown and crisp, remove from the oven and
puncture the cases with the tip of a sharp knife.
Return to the oven at a lower heat to completely dry
out the inside. Cool before freezing.

To freeze: Wrap pastry cases in plastic film and
place in an airtight plastic box for extra protection.

To serve: Thaw pastry cases in wrappings at room temperature for about 1 hour. Fill with whipped cream and sprinkle lightly with icing sugar. Use about 1 pint cream, whipped, and 4-6 tablespoons icing sugar.

PETIT FOURS

Makes: approximately 4 dozen petit fours

1 slab cake 12 × 8 × 1-inches (one day old)
2 oz icing sugar
4 oz almond paste
½ cup apricot jam, sieved
1 quantity glacé icing
decorations as desired

Cut slab cake into shapes such as circles, crescents, triangles, squares, oblongs, no larger than 1½-inches across. Remove all crumbs from working area. Sprinkle bench with a little sifted icing sugar and roll out almond paste to ⅛-inch in thickness. Cut into shapes corresponding to cakes. Any remaining almond paste can be coloured and rolled into shapes to decorate the cake or can be used under the icing to mould the cakes. Coat the top and sides of cakes with hot apricot jam and place almond paste on top. Warm the glacé icing slowly in the top of a double boiler and individually ice each cake, coating it completely. Allow to drip and set on a wire cooling tray. Decorate cakes as desired.

To freeze: Freeze each cake individually for 2-3 hours. Wrap individually in foil and place in airtight containers to prevent damage in freezer.

To serve: Thaw in wrappings for 2 hours, loosening them when necessary to prevent sticking. Serve on dainty plates.

SLAB CAKE

Makes: 1 slab cake (approximately 4 dozen Petit
Fours)
Cooking time: 45-55 minutes
Temperature: 350-375 °F. gas No. 4

8 oz (2 cups) self-raising flour
$\frac{1}{4}$ teaspoon salt
3 eggs
6 oz butter
6 oz (1 cup) castor sugar
1 fl oz milk

Sift flour and salt. Beat eggs. Cream butter and
sugar until light and fluffy. Gradually beat in eggs,
mixing well. Add flour and a little milk alternately.
Fold in thoroughly. Place mixture in a flat oblong
cake tin, 12 x 8 x 1-inches which has been
previously greased and the base lined with
greaseproof paper. Bake in a moderate oven for
45-55 minutes. Cool on a wire cooling tray.

GLACÉ ICING

4 oz granulated sugar
2 fl oz ($\frac{1}{4}$ cup) water
1 tablespoon liquid glucose
12 oz pure icing sugar, sifted

Dissolve granulated sugar in water and boil for 3
minutes. Add glucose. Cool and beat in icing sugar.
Colour and flavour as required. Warm in the top
of a double boiler when required to ice cakes.

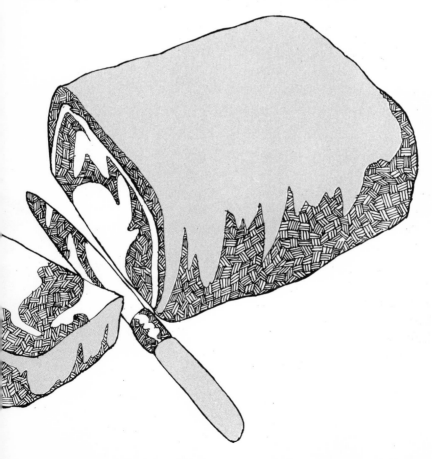

FREEZING FOR A SUMMER BUFFET DINNER PARTY

Summer time conjures up in one's mind an abundance of sunshine, fresh fruit and vegetables and salads, picnics on the beach, lunch in the garden, barbecues, garden parties and patio parties, all of which are easy to prepare when you own a deep freeze. If you have a wedding anniversary or birthday due this summer why not invite your friends and relatives to a buffet dinner party. Preparing an elaborate buffet for a number of people is quite a job, but if you are well organised and use your deep freeze sensibly, it will not be hard work.

Prepare and deep freeze, in advance, a large supply of ice cubes. Once frozen, transfer from ice cube trays into strong plastic or polythene bags, seal and store in your freezer until required. All types of cream cheese dips, croûtes and savoury toppings can be frozen, ready to thaw and assemble quickly to serve with pre-dinner drinks. Fish may be frozen. Meat may be cooked and frozen, all ready to serve with salad. Freeze a supply of bread rolls and butter curls or balls. Make pastry or biscuit crumb flan cases and freeze until required, fill before serving. Make and freeze a gâteau or a fruit dessert. Your deep freeze can help you to help yourself, so be a successful relaxed hostess at your summer buffet dinner party.

Try the following menu for your next summer buffet dinner party. Follow the instructions carefully and serve your guests a cool, refreshing delightful summer meal. The menu and recipes serve 12.

Serve a chilled white wine with the seafood cocktail and a choice of chilled white wine or a rosé wine with the main course.

Canapés
Shrimp Cocktail
Cold Roast Fillet of Beef
Green Salad
Bread Rolls and Butter
Continental Cheesecake
Cheese
Coffee

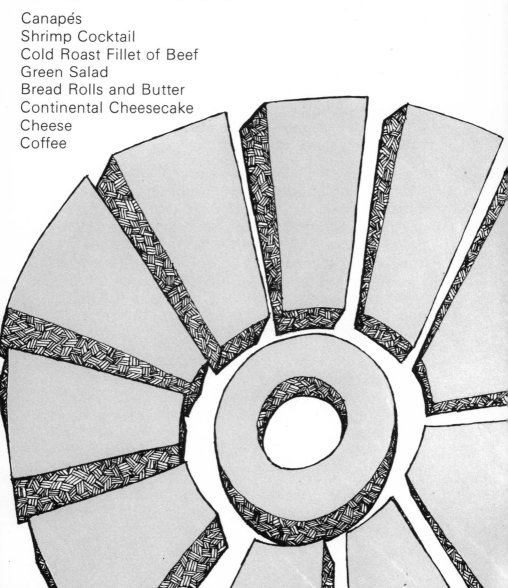

CANAPÉS

Makes: 24

8 slices white bread
oil for frying
Cream cheese topping:
1 × 4 oz packet cream cheese
2 tablespoons cream
salt and pepper
green peas for garnish
Salmon topping:
1 × 3¾ oz can smoked salmon
lemon and capers for garnish

For croûtes: Cut bread into small circles, using a
cutter approximately 1¾-inches in diameter. You can
cut 3 circles from 1 slice of bread. Heat oil, ¼-inch
deep, in a heavy based frypan and fry circles of
bread until golden brown on both sides. Drain well.
For cream cheese topping: Mix cream cheese with
cream until smooth. Season to taste with salt and
pepper.
For salmon topping: Drain salmon well.

To freeze: Package croûtes in a plastic container
or in a plastic bag, seal well, label and freeze.
Package cream cheese mixture in a plastic
container, allowing ½-inch headspace, seal, label
and freeze.
Package smoked salmon flat in a plastic bag, remove
air, seal well, label and freeze.

To serve: Thaw croûtes, cream cheese and salmon
in containers at room temperature. Spread 12

croûtes with cream cheese mixture and garnish with green peas. Cut smoked salmon into 12 neat circles with a cutter, place on remaining croûtes and garnish with thin slices of lemon and capers. Serve canapés on a flat plate.

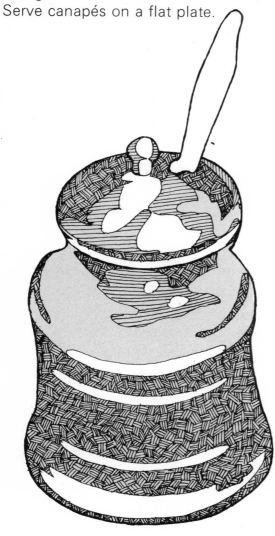

SHRIMP COCKTAIL

Serves: 12

1-1½ lb freshly cooked and peeled or canned shrimps
1 cup mayonnaise or salad dressing
2 tablespoons tomato sauce
2 tablespoons lemon juice
1 tablespoon horseradish sauce or relish
1 teaspoon salt
¼ teaspoon pepper
½ teaspoon paprika pepper
crisp lettuce leaves for serving

Weigh fresh shrimps after peeling. If shrimps are very large, cut them in half. Small prawns may be used in this cocktail if shrimps are not obtainable.

Mix remaining ingredients together except lettuce, until well blended. Taste and adjust flavour if necessary. Fold shrimps into the dressing.

To freeze: Package Shrimp Cocktail in a plastic container allowing ½-inch headspace. Seal well, label and freeze.

To serve: Thaw in refrigerator overnight. Serve chilled in seafood cocktail glasses lined with crisp lettuce leaves or shredded lettuce. Serve garnished with lemon.

ROAST FILLET OF BEEF

Serves: 12
Cooking time: 25 minutes
Temperature: 450-500 °F. gas No. 8-9

1 × 3 lb eye fillet of beef
freshly ground pepper
2 oz butter
1 tablespoon olive oil

Trim fillet of beef into a neat shape, removing all fat
and coarse tissues. Sprinkle beef with freshly
ground pepper and rub into the meat with the
fingers. Melt butter, add olive oil and brush mixture
over beef. Brush a large piece of foil with butter and
oil mixture. Wrap beef securely in foil and place on
a rack in a roasting pan. Place beef in a very hot
oven and cook for 15 minutes. Unwrap foil, fold it
back and continue to cook beef for a further
10 minutes. Cool rapidly.

To freeze: Slice beef into $\frac{1}{4}$-inch slices. Pack beef in
layers with a thin sheet of clear plastic between
each slice. Wrap beef in a sheet of clear plastic,
then place fillet in a strong plastic bag. Remove air
from bag with vacuum pump. Seal carefully, label
and freeze.

To serve: Thaw beef in its wrapping in refrigerator
for 8 hours, or overnight. Serve sliced beef at room
temperature for a better flavour. Serve with green
salad and hot bread rolls.

CONTINENTAL CHEESECAKE

Serves: 8-12

2 oz butter
2 cups chocolate biscuit crumbs
$\frac{1}{4}$ cup sifted icing sugar
$1\frac{1}{2}$ cups cottage cheese
$1\frac{1}{2}$ teaspoons gelatine
2 eggs, separated.
$\frac{1}{4}$ cup cold water
$\frac{1}{2}$ teaspoon vanilla essence
3 tablespoons sugar
$\frac{1}{2}$ cup cream
1 teaspoon grated lemon rind
juice of 1 lemon
extra $\frac{1}{3}$ cup sugar
1 × 15 oz can pears, drained

Soften the butter in a mixing bowl. Using a fork, mix in $1\frac{1}{2}$ cups of biscuit crumbs and the icing sugar. Press mixture evenly over sides and base of a springform cake tin and refrigerate until set.

Place cottage cheese in a mixing bowl and break up into crumbs with a fork. Sprinkle gelatine over cold water and leave to soften.

Combine lightly beaten egg yolks, vanilla essence, sugar and half the cream in the top of a double boiler. Stir over simmering water until slightly thickened. Remove from heat, stir in the softened gelatine. Stir constantly until dissolved.

Pour the mixture over the cheese. Fold in the remaining cream, lightly whipped, lemon rind and juice. Whisk egg whites until stiff, beat in extra sugar and fold into mixture.

A KELVINATOR NO FROST
REFRIGERATOR FREEZER

A SUMMER BUFFET DINNER PARTY
PACKAGED FOR THE FREEZER

THE SUMMER BUFFET
DINNER PARTY SERVED

A KELVINATOR NO FROST
REFRIGERATOR FREEZER

Spoon half the mixture into prepared crumb crust. Place sliced pears on top and cover with remaining cheese mixture. Sprinkle surface with the leftover biscuit crumbs or decorate as desired.

To freeze: Place in a plastic container with a tight fitting lid. Seal, label and freeze.

To serve: When required, thaw cheesecake with opening container, in refrigerator for 6-8 hours. Serve chilled. To cater for 12 guests, you may need 2 cheesecakes, depending on appetites.

 BREAD ROLLS

Serves: 12

12 large or 24 small bread rolls

Fresh bread rolls can be frozen successfully.

To freeze: Package bread rolls in a plastic bag, remove air with a vacuum pump and seal carefully. Label and freeze.

To serve: Place frozen bread rolls into a hot oven (400-450 °F. gas No. 6-7) for 10 minutes. Serve hot rolls buttered, or serve with pats of butter and let guests help themselves.

FREEZING FOR A WINTER BUFFET DINNER PARTY

One of the nicest things about the cold winter days must surely be the reality and appreciation of one's own home. The simple pleasure enjoyed by generations of people sitting comfortably before an open log fire, listening to the wind and rain outside, talking and joking. Another welcome winter pleasure is to give a party for friends, providing them with the opportunity of sharing the comfort and warmth of your home. Providing you make full use of your deep freeze, preparing a buffet dinner party for a crowd will be easy.

Choose a menu first and then prepare and freeze as much as possible, so avoiding a last minute rush. A good hostess should be relaxed and happy to entertain her visitors, not hot, tired and completely 'fagged out'. Even if the meal is delicious, your guests will feel embarrassed by any obvious fatigue caused by providing the meal for them. Be well organised and use your deep freeze to store the party menu. Choose dishes that freeze well and are easy to serve, requiring little last minute attention. Choose dishes that will reheat in a casserole so that they may, if required, be taken from the oven straight to the table.

Try the following menu for your next winter buffet

dinner party. Follow the instructions carefully and serve your guests a delicious, hot meal. The menu and recipes serve 12. Serve beer, cider or chilled white wine with the main course.

Canapés
Chicken and Veal Curry
Sweet and Sour Pork
Boiled Rice
Pineapple Gâteau
Coffee

CRABMEAT PUFFS

Makes: 24

24 croûtes
1 cup crabmeat
½ cup mayonnaise
1 tablespoon lemon juice
salt and pepper
egg white
paprika pepper

Prepare croûtes (see page 84), drain well and allow to cool.
Drain crabmeat and flake with a fork.

To freeze: Package croûtes and crabmeat in separate plastic containers, allowing ½-inch headspace for the crabmeat.

To serve: Thaw croûtes and crabmeat overnight. Mix crabmeat with mayonnaise, lemon juice and season to taste with salt and pepper. Fold in the stiffly beaten egg white and pile mixture on top of croûtes. Sprinkle with paprika pepper and place under a hot grill until puffed and golden brown. Serve immediately, garnished with lemon and parsley.

CHICKEN & VEAL CURRY

Serves: 12
Cooking time: 1 hour

2 chickens
1½ lb veal steak
¾ cup rice flour
6 tablespoons oil
1 green apple, peeled and chopped
1 cup chopped onion
1 tablespoon curry powder
3 teaspoons salt
½ teaspoon ground ginger
¼ teaspoon dry mustard
extra 3 tablespoons rice flour
4 cups chicken stock (use stock cubes)
½ cup cream

Remove chicken flesh from bone and cut into
1-inch pieces. Cut veal into 1-inch pieces. Coat
chicken and veal in rice flour. Heat 3 tablespoons oil
in a large frypan and sauté half the chicken and veal
until golden brown. Transfer browned chicken and
veal to a large heavy saucepan and keep warm over
a very low heat. Heat remaining oil in frypan, sauté
remaining meat until golden and add to first half
in saucepan.

Gently fry chopped apple and onion in remaining
oil in frypan, until soft. Stir in curry powder, salt,
ginger, mustard and extra 3 tablespoons rice flour.
Add stock and bring to the boil, stirring
continuously.

Pour curry sauce over chicken and veal. Stir in cream. Simmer gently over a low heat for 30 minutes. Cool rapidly by standing saucepan in cold water.

To freeze: Place a large thick plastic or polythene bag into a large casserole (to be used as a mould). Pour cold curry into bag, allow to settle in shape of casserole then seal well allowing 2-inch space for expansion. Place casserole into deep freeze until curry is frozen (approximately 4 hours). When frozen, block of curry may be removed from casserole and stored in freezer.

To serve: Remove block of curry from plastic bag and place in original casserole. Thaw for 3-4 hours then reheat in a moderate oven (350-375 °F. gas No. 4) for $1\frac{1}{2}$-2 hours or until heated through. Stir in crushed garlic while reheating. Serve with boiled rice.
Note: It is advisable to store the frozen curry for only 1 month, due to the large amount of onions and the curry powder which could develop 'off flavours' over a longer storage time.

SWEET & SOUR PORK

Serves: 12
Cooking time: 45 minutes

3 lb pork fillet
3 tablespoons soya sauce
1½ tablespoons dry sherry
5 tablespoons plain flour
¼ pint vegetable oil
Sauce:
6 green peppers
2 red peppers
3 onions
8 oz frozen green beans
8 oz frozen 'baby' carrots
6 slices pineapple
¼ pint vegetable oil
3 tablespoons dry sherry
3 tablespoons lemon juice
6 tablespoons tomato sauce
3 tablespoons soya sauce
1 cup sugar
1½ tablespoons rice flour
1½ cups water

For pork: Cut pork into 1-inch cubes and mix with soya sauce and dry sherry. Drain off any excess liquid and toss pork in plain flour. Fry in hot oil until golden brown (about 15 minutes). Remove pork from pan and drain well.

For sauce: Remove seeds from peppers and cut into thin strips. Chop onion, cut beans and carrot into strips and pineapple into cubes. Fry vegetables in

heated oil in a large saucepan until slightly brown.
Mix together the sherry, lemon juice, tomato sauce,
soya sauce and sugar. Add to vegetables, along with
pineapple cubes, and bring to the boil. Blend rice
flour with water to a smooth paste, add to mixture
in pan and bring to the boil stirring continuously.
Cool rapidly.

To freeze: Spread partly cooked pork onto flat trays
to cool. When cold freeze on trays for 2 hours
(free flow method, see page 106). Package in a
plastic bag, remove air, seal well, label and freeze.
Pour cold sauce into a plastic container, allowing
1-inch headspace. Seal well, label and freeze.

To serve: Thaw sauce for 6-8 hours and pork for
2 hours, at room temperature. Heat $\frac{1}{4}$ pint oil in a
large frypan and fry pork until brown and hot.
Drain well and place in a warm serving dish. Bring
sauce to the boil, but do not simmer, and pour
immediately over the hot pork. Serve with boiled
rice.

BOILED RICE

Serves: 12

1 tablespoon salt
2 cups short grain rice

Add $\frac{1}{2}$ tablespoon salt to 8 cups boiling water.
Gradually add 1 cup short grain rice. Boil rapidly,
uncovered, for approximately 12 minutes. Pour rice
into a colander to drain. Repeat the whole process.
Two lots of rice will give you approximately 6 cups
of cooked rice. Once rice is drained, spread out to
cool quickly.

To freeze: Package cold rice in flat pack method
(see page 107) or in a plastic container. Seal
well, label and freeze promptly.

To serve: Break up frozen rice and drop into a large
quantity of boiling water. Immediately rice returns
to the boil drain through a large colander and
serve hot.

PINEAPPLE GÂTEAU

2 large sponge cakes
1 pint cream
1 lb apricot jam
1 × 1 lb 13 oz can pineapple slices
glacé cherries
angelica

Make or buy three 8-inch sponge cakes.

Whip the cream in a large bowl, cover and place in the refrigerator for 3-4 hours. Leaving whipped cream to stand, allows the water content to settle at the bottom of the bowl.

Slice sponge cakes across the centre and fill with whipped cream and jam. Sieve remaining jam, heat gently and brush the tops of the sponges with warm apricot glaze. Decorate the tops with rings of well-drained pineapple. Brush again with the glaze, place glacé cherries and angelica 'leaves' in a pattern on top of each sponge.

To freeze: Freeze cakes in plastic containers, without a cover, for approximately 2 hours. Cover the containers with tight fitting lids, or use biscuit tins as protection and seal well. Label and freeze. Pineapple Gâteau can be stored in your freezer for up to 4 weeks.

To thaw: Simply thaw at room temperature for 3-4 hours. Do not break the container seal during thawing period. Glaze cakes again with warm apricot purée, if necessary, before serving.

IMPORTANT FREEZING POINTS

There are some foods which do not take kindly to storage in the freezer, and so when preparing a programme of preparation and freezing, it is wise to remember the following points.

GARLIC tends to develop an 'off flavour' in casseroles and stews, so add the juice of garlic when reheating these dishes.

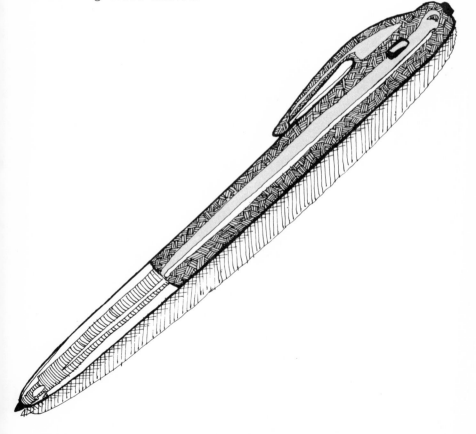

ONIONS in large quantities change flavour when used in prepared dishes which are stored in the freezer for a long time. Short term storage for one month does not seem to alter the flavour, however.

STRONG SPICES in large quantities are not advisable. Allow for extra cooking time on thawing and add them then.

SALT in excessive amounts inhibits the freezing of foods. Do not store ham or bacon in the freezer for a long time.

CLOVES develop a strong flavour, so add them on thawing.

PEAS and POTATOES should not be added to savoury dishes before freezing. Add when reheating.

TOPPINGS WITH CRUMBS AND CHEESE should be added after thawing.

COOKED FISH is not recommended unless it is incorporated in fish cakes, fish pie, etc.

STUFFING should not be frozen in poultry, but can be packaged separately and frozen.

HARD-BOILED EGGS become tough and rubbery.

MAYONNAISE separates out but mixed with other ingredients in small quantities it is satisfactory.

SALAD VEGETABLES such as lettuce, tomatoes, cucumbers, endive, whole onions, celery and radishes lose their crispness and cannot be served in salads. However, tomatoes and celery can be served as cooked vegetables and chopped onions are handy for casseroles.

AVOCADOS do not freeze well, but can be mashed with lemon juice for use in dips.

BANANAS when frozen whole turn brown, but mashed with lemon juice can be used as fillings for sandwiches, etc.

FRIED CROÛTES do not store well unless they are spread with butter.

SALTED CHEESES cannot be served on a platter, but can be used in cooking.

SAUCES with a cornflour base separate out and are not suitable for serving. However, sauces thickened with eggs and cream can be frozen satisfactorily provided the dishes are thawed before reheating in a double saucepan. Rice flour may be used as a thickening agent.

CUSTARDS and caramel creams do not freeze.

GELATINE set jellies loose their bright appearance, and the jelly has an uneven texture. But when set in thin layers between fruit, this texture is acceptable. Gelatine may be used as a setting agent for creams, soufflés, etc.—the effect of the gelatine is masked, and thawed results are good.

SOUR CREAM separates when frozen on its own although may be combined with other ingredients.

FRESH CREAM with less than 40% butter fat separates when frozen on its own.

SOFT MERINGUES go tough and sticky.

VANILLA ESSENCE, if it is synthetic, becomes strong during storage. If this flavour is required, use vanilla pods or vanilla sugar.

PRECOOKED VEGETABLES tend to acquire a 'warmed-up' flavour, but are more acceptable when covered in a sauce.

POWER FAILURE

A well-stocked freezer will keep the contents safely frozen for about 24 hours in the event of a power failure, providing the freezer is kept tightly closed. Dry-ice may be used, and when placed in the freezer with the frozen foods it will keep them frozen in good condition for much longer.

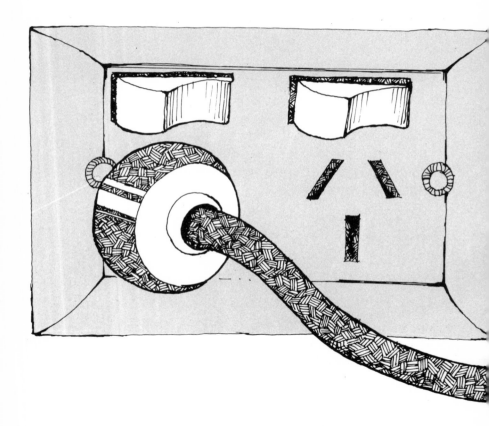

Food which becomes accidentally thawed must not be refrozen. It should be kept at refrigerator temperature, (35-40 °F.) and used within 24 hours, or it should be discarded.

Sometimes an agent or manufacturer will assist with the temporary loan of either freezer storage space, or another freezer, in an emergency.

Allow only experts to check the mechanism of your freezer when something goes wrong. Do not attempt to correct the trouble yourself.

Before sending for the service man check the following points:
(a)   is the main switch on?
(b)   is the electric plug loose or pulled out?
(c)   is the thermostat turned off?
(d)   has the fuse blown out?
(e)   is there a power blackout in the area?

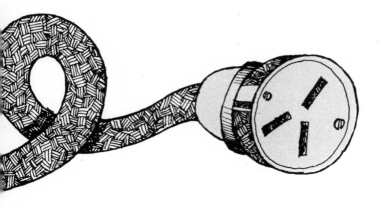

PACKAGING AND LABELLING

PACKAGING MATERIALS
PLASTIC CONTAINERS
An assortment of plastic and polythene containers are manufactured specifically for storing food below zero degrees. They can be purchased from most kitchenware stores. Cheap plastic containers not designed to stand freezing temperatures will become brittle and crack leaving your food unprotected.

Plastic or polythene containers make ideal receptacles for freezing fruit juice, vegetable juice, fruit in syrup, soups, stews, milk, cream and all liquid type foods. These containers are also ideal for sandwiches, cakes, biscuits, party food and leftover foods.

BAGS
Do not use very thin plastic or polythene bags as they can become porous when below zero. Bags specifically made of the correct thickness for deep freezing cost only a fraction extra.

Plastic and polythene bags are very versatile, suitable for freezing practically all types of food. Excellent for vegetables, fruit, meat, fish, sandwiches, cakes, biscuits, savouries and cooked food.

VACUUM PUMP
This is ideal for removing air from plastic and polythene bags. When the pump is placed inside a bag containing food, the air can be removed by

PACKAGING APPLE PUREE
IN ICE CUBE TRAYS AND
FROZEN CUBES OF BABY FOOD
IN PLASTIC BAGS

A WINTER BUFFET DINNER PARTY
PACKAGED FOR THE FREEZER

THE WINTER BUFFET
DINNER PARTY SERVED

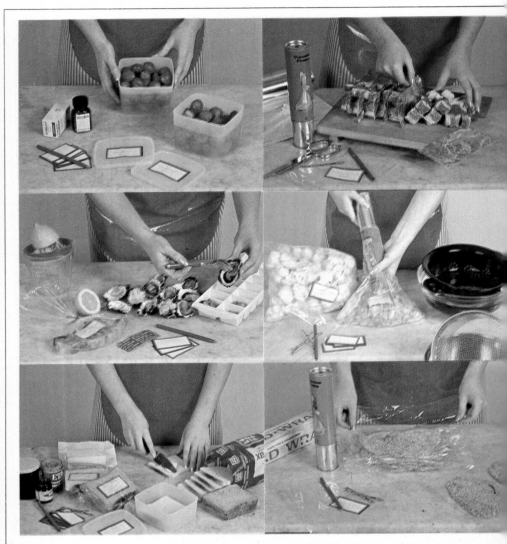

(1) PACKAGING STRAWBERRIES BY
THE RAW FRUIT SYRUP METHOD
(2) PACKAGING FRESH OYSTERS
IN ICE CUBE TRAYS AND
FROZEN CUBES IN A PLASTIC BAG
(3) PREPARING AND
PACKAGING SANDWICHES

(4) PACKAGING LAMB LOIN CHOPS
(5) PACKAGING CAULIFLOWER FLOWERETT
AND CARROT RINGS
(6) PACKAGING VIENNA SCHNITZEL

holding the bag firmly around the outer cylinder, pumping the inner cylinder in and out several times and the air is drawn out of the bag. This simple, but effective gadget can be obtained from leading kitchenware stores.

ALUMINIUM FOIL
Foil is ideal for wrapping meat, fish, cakes, sandwiches, etc. However, it is easily torn or split, so be extra careful when wrapping food with tinfoil.

GLASS JARS AND BOTTLES
Ordinary glass jars and bottles can make suitable containers for deep freezing. They will not break providing you remember to leave 1-2-inches head room for expansion of liquid type foods. If you fill glass jars or bottles to the top, then seal, the below zero temperature will cause the liquid to expand and so shatter the glass.

WAX COATED CARDBOARD TUBS
This type of container is excellent when first new, providing the lids are tight fitting. However, the wax coating is inclined to wear off during the washing. You will find they can still be of use if you place a suitable sized plastic bag inside as a lining.

Tubs are ideal for shell fish, gravy, sauces, fruit and
vegetable juice and all liquid type foods.

TINS

Ordinary household biscuit tins may be used below
zero, providing they are rust free and you remember
to use special freezer tape to seal the lids.

Tins are ideal to protect fragile items such as
meringues and pavlovas etc.—place into an airtight
bag first, then store in a tin. Tins are also ideal
for storing sandwiches, cakes, biscuits and all
baked foods.

LABELLING

The details of what each parcel contains have to
remain legible, even when packaging becomes
dampened by condensation. Stick-on labels can be
written upon with a ball-point pen. A marking pen
will write on all packaging materials but may remain
permanent. A wax crayon will write on many surfaces
and usually washes off with hot soapy water. The
choice is yours.

I strongly recommend labelling all items. You may
think you will remember details of contents but it
is so easy to forget.

PACKAGING METHODS

The method of packaging you choose should depend
on the type of food you are dealing with. All liquid
type foods are naturally best frozen in plastic
containers with tight fitting lids or a plastic bag
sitting in a container as a mould.

FREE FLOW METHOD

To prevent items of food sticking together, I suggest
the free flow method. Simply spread prepared
vegetables, fruit, meat or fish out onto a flat smooth
dry tray. Place the tray without a cover into your

freezer, preferably onto the quick freeze surface, (over the electric motor area). Once the food is frozen it is easily removed from the tray. Tip the loose frozen food into a good quality plastic or polythene bag. The second and third trays of frozen foods can be added to the first and so on until a large bag of loose 'free flow' food is achieved. Remove the air, and seal with a wire twist. The free flow method gives you the convenience of removing individual pieces or portions, as required, from a large quantity without thawing.

FLAT PACK METHOD
Flat shaped parcels stack easily taking up very little space. Simply place meal size quantities into a good quality plastic or polythene bag. Lay the bag flat on its side, shake a little to distribute evenly then gently pat flat. Draw out air with a vacuum pump or carefully press air out with your hands. Seal with a wire twist. Flat packages stack easily which will help keep your freezer tidy.

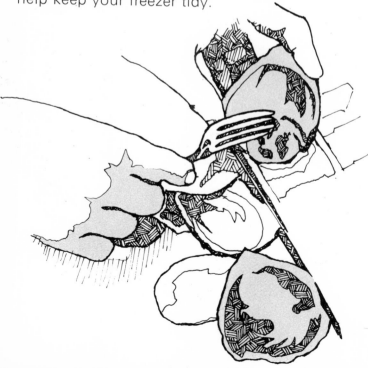

STORAGE TIME GUIDE

The times given below are for your guidance. Food kept longer than recommended will deteriorate in quality, lose its flavour, colour and goodness. Fatty foods can become rancid if kept too long. Poorly packaged foods will deteriorate within weeks, wasting your time, effort and freezer space. I cannot emphasise too greatly the importance of good packaging.

MEAT	APPROX. MONTHS
Meat, lean	12
Meat, fatty	6
Slabs of bacon	6
Sliced bacon	2
Sausages	2
Uncooked ham	6
Cooked ham	4

POULTRY	
Chicken	12
Game birds	12
Turkey, ducks and geese	8

FISH	
White fish	6
Whitebait	6
Shell fish	6
Cooked fish	3
VEGETABLES	12
FRUIT	12

DAIRY FOODS	APPROX. MONTHS
Butter	6
Cheese	6
Pasteurised cream	4
Unpasteurised cream	1
Whipped cream	1
Milk	1

BAKED FOODS	
Bread	9
Cakes	9
Iced cakes	6
Biscuits	9
Biscuit dough	6
Baked and unbaked pies	6
Unfilled éclairs, etc.	2

COOKED FOODS	
Roast Beef	4
Roast Lamb	4
Roast Pork	4
Roast Veal	4
Chicken	4
Casseroles	6
Stews	6
Curries	3
All fried foods	3

THAWING TIME GUIDE

MEAT
All types of meat can be cooked while still frozen.
Always allow 20 minutes per lb extra cooking time.
However, I recommend partially thawing large joints
to allow uniform cooking.
To thaw:
To thaw completely allow, 3-4 hours per lb at room
temperature. Place joint into the refrigerator and
thaw slowly for 5-6 hours per lb.

FISH
Fish will lose a third of its natural flavour and juices
if thawed to room temperature before cooking.
Therefore, I recommend cooking fish from the frozen
state. However, if thawing is necessary do not break
the seal of the package.
To thaw:
Place package in cold water for 1 hour per lb.
Fish thawed in a refrigerator will take approximately
5-6 hours per lb.
Note: To retain the flavour of fish, cook whilst
frozen or partially thawed or whilst still chilled.

VEGETABLES
Do not thaw, always cook from the frozen state.

FRUIT
Can be cooked from the frozen state or eaten raw.
To thaw a 1 pint size quantity (do not break
packaging seal):
(1) Leave package to thaw naturally at room
temperature for approximately 4 hours.

(2) Thaw slowly in your refrigerator for approximately 7 hours.

(3) Place package into a basin of cold water for approximately 1 hour.

(4) Place package before a fan for approximately 1 hour.

(5) If fruit has been stored in the free flow method, you will find thawing time considerably less, 20-30 minutes in a refrigerator.

COOKED FOODS

It is difficult to give an approximate time to thaw cooked foods as it naturally depends on the size and bulk of the item. However, use as a guide the following examples. A meal for 6 is best thawed for 1 hour in the warmth of your kitchen then reheated in the usual way. Allow an extra 30 minutes over the usual heating time. Oven temperature hot (400-450°F. gas No. 6-7).

BAKED FOODS

Thaw in the warmth of your kitchen without removing from the sealed package.

To thaw:

A sponge cake will be ready for filling with fresh cream within 30 minutes.

Heavy fruit cake can take 3-4 hours at room temperature depending on the size.

Dainty fancy cakes at room temperature are ready for eating within 1 hour.

Scones, buns, etc., also take 1 hour at room temperature. However, if needed in a hurry they can be taken straight from the freezer onto a baking sheet and straight into a hot oven (400-450 °F. gas No. 6-7) for approximately 10 minutes.

Savouries with a variety of fillings or toppings can take up to 2 hours thawing at room temperature. However, savouries are usually served hot. Frozen savouries placed onto hot trays in a moderately hot oven (375-400 °F. gas No. 5-6) will be ready to eat within 15-20 minutes.

The average sized pie can be thawed at room temperature within 4 hours. Do not remove the packaging. To bake a thawed pie, place the oven tray into the centre of the oven, preheat to hot (400-450 °F. gas No. 6-7). Place pie onto the hot tray and bake for approximately 25-30 minutes. If the thawed pie was baked prior to freezing, 15-20 minutes reheating time is ample. Frozen pies can be baked straight from the freezer in a preheated hot oven (400-450 °F. gas No. 6-7) for 35-40 minutes. If pie is baked prior to freezing the frozen cooked pie will take 30-35 minutes to reheat.

Sandwiches—do not remove the wrapping. Thawing time depends on the quantity in each package. At room temperature a package measuring approximately 6 x 4-inches will take 3-4 hours to thaw. If thawed in a refrigerator, 6-7 hours. Large quantities are best thawed overnight.

DAIRY FOODS

Always thaw dairy foods in unopened containers in your refrigerator. Allow 5-7 hours per pint or lb size. However, if speed is necessary, stand container in cold running water for 1 hour per pint or lb.

YOUR FROZEN FOOD LOG BOOK

A record of your frozen food is very important. It tells you at a glance what foods you have on hand and where they are stored in the freezer. It helps prevent forgetting a package of food that should not be stored for too long—and it assists in ensuring a constant turnover in the freezer.

A suggested chart is set out on the next page. Each frozen food package should be plainly marked with contents, number of servings and the date when placed in the freezer. Keep your own record of all vegetables, fruits, meats, fish, casseroles, baked goods, etc. You can indicate the location by reference to the shelf (upright freezer) or to the storage baskets—left or right, 'bulk' for the bottom part of the storage section below the baskets, and 'freezing' for the freezing area of the chest type freezer.

Be sure to keep a record of what you remove from the freezer, then you will know how many packages remain.

Type of food	Package Size	No. of Packages	Date Frozen	Location	Package Remove

pe of food	Package Size	No. of Packages	Date Frozen	Location	Packages Removed

Type of food	Package Size	No. of Packages	Date Frozen	Location	Package Removed

ype of food	Package Size	No. of Packages	Date Frozen	Location	Packages Removed

ACKNOWLEDGEMENTS

G.E.C. would like to thank the following for
their help and co-operation in the preparation of
this book:
Pauline Holden, Consultant Home Economist for
some of the recipes in this book

Comalco Ltd., for supplying props for photography
Corning Glass Works, for supplying props for
photography
Hav-V-Sales Tupperware, for supplying tupperware
for photography
Incorporated Agencies Pty. Ltd. for supplying china
and glass for photography
Josiah Wedgwood & Sons (Aust) Pty. Ltd. for
supplying china for photography
and the following people for supplying equipment
for kitchen settings in the photographs in this book:
G. Cronk & Son (copper hood)
Formica Plastics Pty. Ltd. (formica bench top)
Kenwood Peerless Pty. Ltd. (electric mixer)
Monarch Kitchens Pty. Ltd. (cupboards)
Prestige Pty. Ltd. (cupboard knobs)